INTER-CITY 125

High Speed Train (1972 onwards – all models)

First published in June 2019

A catalogue record for this book is available from the British Library.

ISBN 978 1 78521 266 6

Library of Congress control no. 2018967907

Published by Haynes Publishing,
Sparkford, Yeovil,
Somerset BA22 7JJ, UK.
Tel: 01963 440635
Int. tel: +44 1963 440635
Website: www.haynes.com

Haynes North America Inc.,
859 Lawrence Drive, Newbury Park,
California 91320, USA.

Printed in Malaysia.

Acknowledgements

The author would like to extend a very special thanks to Dave Moore, Senior Design Engineer, Wabtec for his assistance with this manual. We would also like to thank the following for assistance with material included: Paul Biggs, Jack Boskett, Phil Buck, Paul Colbeck, Paul Corrie, Kevin Daniel, Sam Dixon, Steve Hampton, Doug Harvey, Steve Hoather, Chris Hopkins, Roy Kethro, Martin Loader, Mark Parsons, Cliff Perry, Roger Senior, Peter Sharp, Martyn Spencer, Peter Starks, Peter Swift, John Tattersall, James Trebinski, Lucas Walpole, Bill Watson, Chris Weatherall and Nigel Yule. Also, Abellio, Angel Trains, Porterbrook, ScotRail and Wabtec.

This book was co-ordinated and edited on behalf of 125 Group by Chris Martin.

INTER-CITY 125

High Speed Train (1972 onwards – all models)

Owners' Workshop Manual

An insight into the design, construction, operation and
maintenance of the classic passenger train

125 Group

Contents

LEFT No 43093 *Lady in Red* was one of the first recipients of the bright red Virgin Cross Country livery. It is seen in fine, early morning light as it climbs the 1-in-37 Lickey Incline while leading 1E28, the 07.04 Bristol Temple Meads to Newcastle on 24 May 1997. The location is Pikes Pool, approximately halfway up the formidable gradient, which is well within the capabilities of the IC125, provided both power cars are working correctly! *(Martin Loader)*

Foreword

by Gary Heelas, Chairman, 125 Group

Many happy hours were whiled away during my younger days watching the mighty Inter-City 125s departing from stations, doing exactly what they were built for: powering trains full of passengers along Britain's fastest main lines. This was at levels of comfort and speed which set new standards at the time, and in retrospect, have yet to be beaten in this country. It was clear to me this was an engineering masterpiece. The Paxman Valenta engine increased in speed in reaction to the demand for more power. This was followed by the almost deafening scream from the single turbocharger as the train was lifted seemingly effortlessly from a standstill.

The rear power car would be doing the same, sending small children and those of a nervous disposition running for cover, or at the very least clasping their hands to their ears. But not me! At the time it seemed most other enthusiasts were disinterested in the 125, perhaps as part of the British psyche to champion the underdog or perhaps because the 125 was seen as just a predictable, reliable staple of the railway scene and therefore uninteresting. To me, the 125 was more like a sleek, winning greyhound; very much the top dog in terms of railway traction!

The popular railway press at the time pretty much ignored the 125 – contemporary magazines were largely written and edited by enthusiasts of earlier types of traction which had been replaced or demoted when the 125s were introduced. A mixture of bitterness, indifference and resentment resulted in 125s being ignored, or only featured when something had gone awry. Two of us started the 125 Group in early 1994 to right that wrong. The aim was to provide a focal point for information and news on our favourite train and engage with other followers of the type, which we felt sure must exist and feel equally ostracised by the mainstream railway media.

A long term aspiration would be the preservation of a handful of vehicles once their service careers were over. There was often talk of replacing the 125, but the indispensable nature of the train meant this would seemingly never happen. Even when the type was replaced on one route they soon found themselves redeployed on other lines making preservation always seem a distant goal. Right from the beginning, the pages of 125 Group magazines were filled with news from the

ever-changing scene as the trains were given new liveries, new routes to work and, crucially, re-engineered to allow them to continue in service for such an incredibly long career.

This was to present 125 Group with a new challenge and take us into hitherto uncharted territory. With fleet-wide re-engineering it was realised major components such as engines and radiators would need to be purchased and stored so that our aspiration to return vehicles to their original configuration could ultimately be achieved. From humble beginnings, 125 Group had grown and crucially, not all of our supporters were ordinary railway enthusiasts. We had attracted the attention of senior railway figures who had spent their careers operating the trains, which they knew from first-hand experience, were something rather special.

With vital parts now secured it felt like we may lose some impetus until the long-term goal could be realised by the eventual withdrawal of the fleet. But in 2011, a whole new challenge came along. The restoration of the prototype power car (No 41001) after almost three decades as a static museum piece placed the group at the bottom of a steep learning curve. There were many hurdles on the way to the unforgettable day in November 2014 when it successfully hauled a train full of passengers. As 125 Group enters its 25th year and surpasses the 600-member mark we are finally within sight of being able to secure a production power car. No doubt there will be days in the future when things fail to go to plan, but the end result will be a secure future for an example of the most successful diesel passenger train ever created.

BELOW No 43070 leads the 'Wessex Scot' – the 07.10 Edinburgh to Poole – through Pangbourne cutting on 28 April 1993. Great Western services normally use the lines in the foreground, but generally the inter-regional trains joined the relief lines as seen here. Seven carriage formations were the norm for cross-country workings as lower demand for First Class accommodation meant one such carriage was sufficient. *(Martin Loader)*

Introduction

Much has been written about how the Inter-City 125 was the train that revived fortunes on British Rail's long-distance routes during the late 1970s, but it has turned out to be so much more than that. True, the first thing that the IC125 did was to reverse the declining trend and win back passengers who had taken advantage of the expanding motorway network, with its convenient private cars and cheap coach services.

Raising demand by one-third in the first three years of operation on the London to Bristol and South Wales routes was just the first of this train's many conquests. Increasing airline competition on the London to Edinburgh route was soon answered by the new, much faster trains, but it wasn't always about top speed. When introduced, the IC125 generally replaced 90mph trains (apart from a handful of crack expresses in the hands of 100mph locomotives), but they also heralded a leap forward in passenger comfort.

Although the focus of this book is on the IC125 power cars, it was the carriages marshalled between them where the passenger noticed the biggest changes. Fully air-conditioned, sound-proofed and smooth riding, the new Mark 3 carriages were revolutionary when compared with the stock they replaced, particularly on the Western Region lines. There had never before been such a stark improvement in passenger accommodation, and even on lines where comparatively modest journey time gains were achieved, the new trains rejuvenated demand.

Brand-new sleek trains were the marketing department's dream and the impact of the prestigious IC125 spread far wider than the routes they actually served. They also brought about social change: the faster journey times gave rise to the long-distance commuter. It was now possible to live in rural Wiltshire or Lincolnshire, drive to a main line station and reach London in around an hour, thanks to the IC125.

Something for which IC125s will always

be remembered is the longevity of the type. It was to be a full 43 years in demanding front-line service on prime routes before any examples were replaced and withdrawn without prospect of finding further useful employment. The respected railway journalist Roger Ford summed it up perfectly: "No train has been so good for so long." The IC125 had become the train they seemingly could not replace; the train against which others were judged. Even when replacements have arrived they have come with a top speed no higher than the 125mph introduced to passengers in 1976. They have typically featured noisy, underfloor engines which intrude on passenger comfort and they have always been more expensive to purchase and operate than the IC125.

Even when the type disappears from its original stamping ground on the main lines from London Paddington and King's Cross, numerous examples will remain in operation, with a reduced number of carriages, on secondary routes. This will see some of them still operational after more than 50 years of daily service. They will be sorely missed by many passengers, staff and railway authorities on the routes no longer covered.

Of course, other countries have fast, sleek trains and many of those operate at far greater

speeds, but the sheer genius of the IC125 was its ability to operate on existing lines amid existing trains, avoiding the need to build expensive (and often bitterly opposed) new lines, or to find the considerable funds needed to undertake extensive electrification schemes. Former Inter-City director Dr John Prideaux described the IC125 in relation to the French high-speed services by calling them "an outstanding achievement, fully the equal of the TGV".

One of the key attributes of the IC125 was that they were not an exclusive luxury for the elite. There were no premium fares to travel on them, and their domination of the routes they worked meant all services benefited rather than just a small number of expresses.

Such a long service career would not have been possible without considerable effort from those responsible for maintaining them. Ultimately the introduction of new power units and other major component replacement, which took place in the 2000s, have kept the fleet fit for purpose. Reliability across the board peaked between 2013 and 2017 when at least three of the five IC125 operators managed to run over 12,000 miles between technical incidents, defined as a fault resulting in a delay of three minutes or more.

A concerted effort by Craigentinny depot in Scotland raised its fleet on the East Coast Main Line to an average of nearly 18,300 miles per technical incident over the same period. You would need to have been very unlucky indeed to have been a passenger on a broken-down IC125 during that period, especially considering the redundancy of having two power cars which would almost always keep the train going even after the most serious technical incident, albeit at a reduced speed.

Now, as the heyday of the IC125 reaches its conclusion, thoughts turn to preserving at least some examples of the type and restoring them to their original condition, to remind us of this glorious period in UK high-speed travel and for future generations to enjoy. The trains have always been firm favourites amongst operating staff, ranging from drivers to engineers, but they have generally failed to gain the recognition they deserve from the railway enthusiast community.

Thankfully, the dedication and commitment of a small band of enthusiasts (with the generous support of many key figures in the rail industry) has enabled the production of this manual, the restoration of the prototype power car to operational condition after more than three decades in a museum, and will soon enable the restoration of some of the vehicles which have conveyed countless millions of passengers during their long careers.

ABOVE A pair of back-to-back ScotRail power cars are seen on their former home territory on 10 August 2018. Nos 43140 and 43149 ran from Brush, Loughborough to Laira depot, Plymouth for attention, creating a rare visit for the 'Inter7City' ScotRail livery scheme, seen here in Bristol.
(Lucas Walpole)

Chapter One

The need for speed

New, faster passenger trains were needed for the railway to remain competitive. Whilst the cutting edge technology for the Advanced Passenger Train was developed, a simpler solution with a shorter lead time was called for. Enter the Inter-City 125 – the train that was to reverse declining demand and become a frontline icon for over four decades.

OPPOSITE The 07.52 Milford Haven to Paddington is just arriving at Didcot Parkway on 26 March 1986 powered by 43138 with 43134 at the rear. Such was the success of the Inter-City 125 service that stations with large car parks became popular with commuters, now able to live in rural areas and reach London quickly and easily. *(Kevin Daniel)*

Growth in the availability of the private car and improvements to the road network during the 1960s heralded an end to the dominance of the railway for medium and longer distance journeys throughout the UK. British Rail (BR) had gone through a modernisation plan beginning in the previous decade, but this had largely focussed on replacing steam with cleaner and less labour-intensive diesel trains, without delivering a leap forward in terms of train speeds. Investing in new, faster trains was considered essential to maintain a competitive edge.

The Railway Technical Centre (RTC) in Derby was built and in 1962 Dr Sydney Jones was appointed as the new Director of Research. By 1966 the Advanced Projects Group was established to develop new passenger trains. During 1967 the basic concept of a new Advanced Passenger Train (APT) was announced, which was to borrow as much technology from airline development as it did from traditional railway vehicles. APT was to operate much faster over existing alignments whilst still maintaining passenger comfort by tilting carriages into curves. Approval to build the gas turbine powered prototype came in 1968.

There was still a need to develop conventional railway technology, so in 1967 newly appointed Director of Design, Walter Jowett, produced studies looking at future requirements. Amid these documents was the concept of a diesel-electric multiple unit train: a Bo-Bo wheel arrangement locomotive marshalled at either end of ten coaches formed of the new disc-braked successor to the Mk2 carriages. Each locomotive was to be powered by a 3,000hp 16-cylinder Paxman Mk3 Ventura diesel engine which would be capable of operating the train at 150mph.

A fixed formation higher speed diesel multiple unit was not an entirely new idea as five 90mph *Blue Pullman* sets had been built as luxury express trains in 1960. These featured a small passenger seating section in the rear of each driving car with the engines located between this area and the driving cab. The design was not successful and the type was withdrawn in 1973, but they did prove the viability of a fixed-formation train with a driving cab at both ends and power provided simultaneously from two full-sized engines.

In mid-1968, a new Chief Engineer for Traction and Rolling stock was appointed. His name was Terry Miller and his role was to be crucial in the creation of the new train. He had trained under Sir Nigel Gresley in the era when fast, steam-powered trains operated along the East Coast Main Line. In his later role as Chief Engineer on the Eastern Region he had overseen the introduction of the Class 55 'Deltic' fleet – a high-speed, lightweight passenger locomotive which had borrowed engine technology from marine applications.

The new diesel-electric multiple unit train was to have its wings clipped slightly before it reached the drawing board. A new, lower maximum speed of 125mph was deemed possible without modification to existing track and signalling systems and this would significantly contain the cost of introducing the new train. Whilst the APT would use its ground-breaking technology to overcome such limitations, the new diesel-electric multiple unit was always destined to be a concept which gained merit by virtue of being cheaper, easier and quicker to develop. The politics of developing a second 150mph passenger train would have been awkward at this time, as BR's aspirations were primarily focussed on the APT, so any train capable of the same speed would have been seen as a direct rival. It was also decided that seven of the new Mk3 coaches would be sufficient and to power the train two engines each capable of 2,250hp would be satisfactory.

Paxman had renamed their Mk3 Ventura diesel engine as simply 'Valenta' and this engine would be capable of being uprated to 2,500hp allowing two locomotives to handle an eight-carriage train should it be deemed necessary. In an earlier trial two Venturas rated at 1,135hp had been fitted to a Class 42 diesel-hydraulic loco but in order to get that much power from a comparatively small engine a considerably larger turbocharger would be needed and each cylinder would require its own fuel pump.

Each of these new locomotives would feature a single driving cab with an aerodynamic front end profile. Crucial to the ability to run at 125mph was the ability to stop the train within existing signal spacing and the new train would feature newly developed disc brakes

and electronic brake control to facilitate this. Developments in track construction, such as concrete sleepers and continuously welded rails, enabled the new train to safely operate at 125mph, provided axle loadings did not exceed the 17t per axle as achieved by the Class 55 'Deltic' locomotives. Dividing the power output between two smaller locomotives at either end of the train allowed this axle loading limit to be achieved. It also meant that in normal use the train would stay in a fixed formation without the need to couple and uncouple locomotives at terminal stations.

Concern over the ability of the research department to develop the APT in an acceptable timescale was seized upon by Terry Miller and his team, and in February 1969 they made a formal submission to the BR board to develop the new diesel train. Perhaps somewhat cheekily, they promised a prototype train of two power cars and eight carriages ready for testing, just 22 months after approval, capitalising on the long lead time needed to create the APT.

Behind the scenes some of the work on the new diesel train was already underway despite no firm commitment having been received to fund the project. Design work on the new Mk3 carriages had progressed. The new coach was to be three metres longer than its predecessor at 23m and new BT10 bogies with a design speed of 125mph were developed featuring air suspension, disc brakes and a wheel slide protection system. Simultaneously, the Royal Navy had conducted extensive type testing on the new 12-cylinder Paxman Valenta engine ahead of installation in military vessels. In August 1970, the BR Board's Investment Committee authorised the expenditure of £800,000 on the construction of an eight-vehicle prototype train, and the real work then began. The new train was to be known as the High Speed Diesel Train (HSDT). With two teams now working on separate fast passenger train designs some rivalry was inevitable. It was very much a clash of old school experience-born railway engineering versus cutting-edge science, and this provided impetus to be the first to turn out a working train.

The power car cab shape was created by industrial designer Kenneth Grange. He had initially been appointed to design a livery for the new train, but not content with that he soon found himself tinkering with models and using a wind tunnel at Imperial College London, developing a more streamlined and better-looking shape for the front of the new train. Although streamlining was nothing new, the importance of applying the science behind it in railway applications was being understood. The clean lines of the Mk3 carriage skirt and the presence of a nose cone-fitted locomotive at the rear of the train, both contributed towards a reduction in wind resistance.

ABOVE Great Western Railway is retaining a small fleet of HSTs to operate Cardiff to Penzance services with four-carriage formations using stock modified with sliding doors. The 12.57 Penzance to Newton Abbot is seen crossing Nottar Viaduct as it spans the River Lynher near St Germans on 20 July 2018. The train is powered by No 43198 *Oxfordshire 2007* and No 43187. *(Chris Hopkins)*

Although the prototype HSDT emerged a month after the APT-E (experimental) in June 1972, the difference between the two trains could not have been starker. The APT-E was simply a test bed while the HSDT was a fully equipped, finished train ready to carry passengers. Terry Miller retired in 1973 so the testing and development phase was overseen by Sid Burdon after his appointment as Locomotive Design Engineer/HST Project Engineer, and Bruce Sephton from the Traction Design Office.

The two prototype locomotives were built at Crewe Works and numbered 41001 and 41002, and were moved to Derby for completion, testing and marshalling into a train. At this stage the train was regarded as a pair of locomotives marshalled either end of a set of individual locomotive-hauled trailer vehicles numbered in the coaching stock series. Ten prototype Mk3 coaches were constructed at Derby Carriage Works. The extra length of the Mk3s increased seating capacity and they featured tapered doors at the carriage ends allowing them to negotiate tighter radius curves without exceeding the loading gauge. As they were configured as conventional locomotive-hauled coaches each was fitted with end buffers and drop-head buckeye couplings. The auxiliaries were powered by motor-alternator sets fed from an 850V DC supply from one of the power cars. All vehicles in the prototype train were turned

BELOW Prototype power car No 41001 on 2 July 2017 while working in preservation. *(John Zabernik)*

out in the then 'Pullman' style livery of pale grey with a dark blue band running the length of the train around the passenger window line.

The locomotives worked with each other in tandem by means of a 36-way cable running the length of the train. By energising and de-energising wires within this cable, control signals were sent between the two locomotives. This system is explained in greater detail in Chapter Two.

The front end of the two prototype locomotives marked a departure in design. Newly developed Glass Reinforced Plastic (GRP) moulded cabs provided a smooth, clean, streamlined appearance. The horns were incorporated above the cab windscreen and the driver was positioned centrally behind a single-piece windscreen. This was produced from newly developed armoured glass to protect the driver in case of objects maliciously or accidentally hitting the screen. Below the windscreen was a second screen covering the centrally positioned tail-light (top) and headlight (below), and either side of the headlight two marker lights were provided. The drawgear was hidden behind a cover which would lift out of place when access to the coupling gear and pipes was required. Conventional buffers were provided, but these were encased in the streamlined cab moulding and enabled the HSDT to be attached to any locomotive in the event of failure.

At the heart of both HSDT locomotives was the newly developed 2,250hp Paxman Valenta RP200L diesel engine. This engine features a compact 60° V-formation, making it ideal for the restrictive loading gauge available within UK rail vehicles. At full speed, it runs at 1,500rpm. The cylinder bore is 7¾in with a stroke of 8½in.

The Valenta had seen significant improvements compared with its predecessor Ventura design with cylinder heads, con rods, pistons and bearings redesigned to cope with the higher firing pressure. The crankshaft was enlarged resulting in a stiffer shaft better able to withstand the harder workload now expected. The engine had been designed using imperial measurements, but upon BR's insistence it was reworked to operate on metric measurements, this itself created some inconvenience for maintainers hitherto used to working on engines using imperial tooling.

The whole engine was constructed from a cast iron block with aluminium ancillary items including the turbocharger, to reduce overall weight. Attached to the engine (but located in the adjacent clean-air compartment) is a Brush BL 91-32 main alternator. This provided traction power of 1,480kW at 1,500V for the propulsion of the train, coupled to a dual-wound Brush BL91-20 auxiliary alternator providing power of 420kW at 850V DC for the carriages. A separate winding on the auxiliary alternator provided 33kW at 110V to feed systems on the power car such as battery charging and lighting. AC supply produced by the alternator is converted to DC by passing through rectifiers located above the alternator set.

Located above the engine was a Napier SA-084 turbocharger, a large single-turbine turbo adding a full 900hp to the engine, which would have been rated at 1,350hp if normally aspirated. Thankfully, any early thoughts of equipping HSDT with gas turbines had long since been abandoned as global politics in the 1970s sent the price of oil rocketing, such that even a comparatively frugal train with two diesel engines looked like an extravagance.

The Marston Excelsior cooler group made best use of the restricted space by featuring two vertical lower panels to provide cooling for the oil cooler, turbocharger and exhaust manifold and two upper panels housed at 45° angles above, to provide cooling for the engine. In the centre of the cooler group is a large fan driven by a gearbox off the free end of the engine, with a thermostatically controlled hydraulic clutch.

The bogie of the HSDT incorporated much of the new design for the train. Rather than turning a wheel in the cab for the parking brake this was applied and released using air pressure. Two four-pole Brush TMH68-46 traction motors were bogie frame suspended, Girling disc brakes were fitted to all axles, and electronic wheelslip and wheel slide protection was fitted. If wheelslip is detected power is removed and reapplied. If wheel slide is detected during braking the system will automatically rapidly release and reapply brake pressure until the wheel slide is corrected.

ABOVE Running at speed on the Great Western Main Line, No 43142 is leading the 08.43 Swansea to Paddington on 24 February 1986. Re-liveried No 43126 can just be seen on the rear. The inadequacies of the original sun blinds meant that supplementing them with a bit of paper or cloth was common practice. *(Kevin Daniel)*

controls for the Electric Train Supply (ETS), the windscreen wiper and parking brake along with the duplex gauge showing the main reservoir supply and main reservoir pipe pressures.

The key controls are located in front of the driver below the windscreen. On the left is a duplex gauge for the bogie brakes, each of the bogies having a separate independent brake system and next to that is the brake pipe gauge. Below the bogie brake gauge are two buttons which operate the brake overcharge (for the purpose of releasing dragging brakes), and a fire alarm test button which has the additional function of shutting down the local engine. Below the brake gauge are two fault indicators, the left one illuminates red in the event of an AWS defect and the right one is a general fault alarm which illuminates red whenever a loss of power or other critical fault occurs on either engine.

On the horizontal surface to the left of the driver is an emergency brake plunger and the seven-step brake controller. Unlike earlier locomotives, a single brake controller is provided rather than separate controllers for the whole train and just the locomotive. To the right is the speedometer capable of showing speeds up to 150mph but with a marker at the 125mph design limit of the train. Beside that is the ammeter

ABOVE Cab interior, left.

BELOW Cab interior, centre.

On the left-hand side of the cab interior is a switch panel which controls the frontal and internal lighting. Nearest the windscreen is the Automatic Warning System (AWS) indicator which shows the driver the status of the previous signal. Below this panel are the

LEFT Cab interior, right.

BELOW The general layout of the cab can be seen viewed from outside. Note the single, centrally located driver's seat which was to be the cause of industrial unrest and delay in testing the new train. A tip-up seat is mounted on the rear bulkhead wall, over the driver's right shoulder. The only ventilation provided for the cab of the HSDT was the provision of drop-light windows in the cab doors. Not everything is as it would have been in 1972; some items have been added to enable this vehicle to meet modern comfort and safety equipment requirements.

showing the power unit output from zero to 2,000 amps. Below the speedometer are two further fault lights, on the left the wheelslip indicator and on the right the engine stopped indicator. It should be remembered that faults on the other locomotive at the rear of the train would be less readily apparent to the driver. Below the ammeter are the buttons for engines stop and start which control both power car engines. On the horizontal surface to the right of the driver is the five-notch power controller and beside that the master switch and associated key socket.

Nearest the driver is the AWS reset button (missing on this vehicle at time of photography) and finally on this panel is the horn; a two-position high/soft tone horn was provided. Out of view on the floor below the windscreen is the Drivers Safety Device pedal, initially this would have just been a 'dead man's' pedal to which the driver must apply pressure to avoid an automated brake application. This was later modified to include a vigilance device requiring the driver to release and re-apply pressure on the pedal in response to a buzzer every 60 seconds or so. Much has changed on the right-hand side wall. The digital clock and an isolation switch are both retrofitted items. Below these was the original analogue clock position,

now replaced by a warning sounder. There would then have been a dimmer switch for the illumination of the cab dials and a variable dial control for cab heating. A buzzer and telephone handset for communication to the guard are still present on the right-hand edge.

ABOVE The rear of the prototype HSDT included a guard's compartment and a small secondary driving position. Although incomplete at the time of photography, the desk originally consisted of a brake controller, a horn, a windscreen wiper controller and a three-position power switch. This switch allowed power notches one and two only to be selected. The cab was intended only for low-speed shunting purposes and was designed to have been used when marshalling trains together.

BELOW External rear view of the prototype. It was never intended to haul trains from this end, but in subsequent preservation anything is possible. This view also shows the luggage van door arrangement with two hinged doors. (John Tattersall)

The single seat cab layout was a new departure, all previous locomotives having side-by-side seating providing for a driver and second man. This was to prove to be a major sticking point as the drivers' trade union ASLEF was insistent both staff should have an equal view of the track ahead, and was determined the new trains would not pave the way to one-man operation. Inflaming the situation further, they sought enhanced payment for their members operating the new, faster trains. The trade union blacked both the HSDT and the similarly configured APT-E and it was not until June 1973 that high speed trials of the HSDT began.

Initially based at Neville Hill, Leeds, it was operated to Newcastle and back along the fastest section of the East Coast Main Line. Although the new Mk3 passenger coaches were found to be riding very well with their new secondary air suspension and insulation creating a very pleasant passenger environment, all was not well with the riding on the locomotives. Severe bogie hunting was experienced at full speed and by 128mph this was intolerable. The solution was to increase the stiffness of the dampers and exchange the flexible mountings between the traction motors

and bogie frame for solid mountings. In an effort to minimise track wear the bogies had been designed to contain the absolute minimum of unsprung equipment and much of the early test running was focussed around tweaking the bogies to obtain the smoothest possible ride from the locomotives.

Once modified, the HSDT was formed as a reduced five-carriage set to carry out high-speed trials and it was during these, on 12 June 1973, that the set reached 143.2mph, a new world speed record for diesel traction.

Following testing, a modification was needed to the carriages. The new disc brakes gave off an acrid burning smell when braking, which would be drawn into the passenger saloons by the air-conditioning. Changes were made which closed the air intakes during braking.

Ongoing union resistance to the new train, and in particular their reluctance to work on it without receiving an enhanced rate of pay, prevented the train from entering passenger service, so plans to operate a regular Leeds to Edinburgh service were shelved.

Between June and October 1974, the HDST was given a 100,000-mile overhaul at Derby Works and during this attention

some changes were made. First, the train was reclassified and instead of being a pair of locomotives with separate carriages it was to be regarded as a fixed-formation diesel multiple-unit. The locomotives were subsequently to be known as power cars and were renumbered with 41001 and 41002 becoming 43000 and 43001 respectively.

The trailer vehicles were also renumbered from the locomotive-hauled series into a multiple-unit series. Two vehicles were removed to be converted into Royal Train vehicles, reducing the train to a regular eight-carriage set.

ABOVE Terry Miller was honoured for his role in developing the most successful passenger train in the UK when No 43048 received the name *T.C.B. Miller MBE* in 2008.

LEFT Set No 252001 is seen near Didcot during passenger service trials while working the 13.48 Weston-super-Mare to Paddington service on 2 July 1975.
(Robin Patrick)

The whole train was classified as No 252001 and this number was applied to the front of both power cars. The designation as Class 252 slotted the HSDT into the diesel-electric multiple-unit sequence after the now-withdrawn Class 251 *Blue Pullman* sets.

During the overhaul, some technical changes were made, the most significant of these being a modification to the alternator on No 43000. The brushless AC alternator on the power cars made it suitable for feeding electric power for the train auxiliaries directly from the auxiliary alternator rather than converting it to DC, sending it down the train to motor-alternator sets under each carriage, which then converted the DC supply to AC power for the carriage equipment. This change could eliminate the weight and unreliability of individual motor-alternator equipment under each coach.

No 43001 received BP10A bogies with Clouth suspension and was subjected to three weeks of static testing in the APT building at Derby before release back into traffic. This power car also received modified engine air intake louvres during the overhaul.

In early December 1974, No 252001 was transferred to Old Oak Common, West London to allow the new train to enter passenger service on the Western Region which was to be the first recipient of the production version of the train. Following a period of staff training and route clearance work, No 252001 entered passenger service on a regular weekday diagram from 5 May 1975, initially at a maximum speed of 100mph. Other testing and maintenance took place at weekends. The set remained in service until summer 1976 and it made a brief return to passenger use to supplement the production sets for the first few weeks of operation in October 1976 until enough production trains were ready to enter service and the non-standard Class 252 could be withdrawn.

The first production order was placed in April 1974 for 27 of the new trains, now known as HST; the 'D' for diesel used in the prototype being dropped. At this time, BR's press department was still sticking to the notion that the APT would be the train of the future, so it was announced that the new trains would only operate the prestigious Inter-City routes until APT was introduced. After that, they would be cascaded onto secondary routes enabling those to be upgraded.

The gas turbines powering the APT-E had proved to be far too thirsty, resulting in a switch to AC electric power for the passenger prototypes. The HST became known as simply being a 'stop-gap' solution while the APT was developed for primary Inter-City routes – that moniker sticking with the HST for its whole life.

In reality, by the mid-1970s APT looked to be only viable for the already electrified West Coast Main Line where speed gains from using tilting trains were also significant. It was quickly realised that no business case could ever be made for replacing nearly new HSTs with APTs on routes where expensive electrification would also be required, and the journey time reduction less significant.

The ongoing issue with the cab layout rumbled on and appeasing the trade union was the only way forward for the production model of the new train. In order to achieve this, a substantial redesign of the cab was necessary. BR went back to Kenneth Grange and he re-worked the design in order to accommodate the requirement for the two drivers to sit beside each other. A larger windscreen would be required. Fortunately, advances in glass production brought the ability to produce a bigger screen still capable of resisting the same amount of impact force. In answer to another of the union's gripes, side windows were provided and with all this extra glass it was soon realised that air-conditioning would be required in the cab to combat the build-up of heat in summer months. It was decided that the new train would not require conventional buffers as it would normally operate as a fixed-formation train. This allowed Grange to come up with a more raked-back front design with the angled windscreen and nose cone shape that gained universal admiration.

As with the prototype, the drawgear, cable sockets and air/brake pipes were hidden behind a hinged cover, but on production power cars the front valance continued further down concealing the front of the bogie. Grange had created a new, sleek and modern front nose cone for the train which became instantly recognisable and was to become an icon of design. The flush appearance of the front end was completed by the relocation of the headlights, marker lights and horn grille into a neater line just above the point of the nose. The first few power cars emerged in a black and yellow livery but before entering service the black areas were repainted into BR's ubiquitous 'Rail blue' resulting in the power car livery matching the carriages.

Experience with the prototype resulted in a few further technical changes on the production model, but these were surprisingly limited. A combination of lack of time and money to update designs, the shorter than expected trial period following the union problems, and simply the skill of Terry Miller and his team in getting the train 'right first time' meant changes were minor.

The cooler group was slightly enlarged and moved closer to the engine while the traction motors were to be cooled by traction motor blowers rather than relying solely on internal motor shaft fans. In the rear of the power car the second driving position was omitted as it was expected the trains would largely operate in fixed formation with no need to attach or detach power cars except for major attention.

The luggage van area was redesigned so a flat floor allowed large parcel trolleys to be wheeled into the area and the pair of hinged luggage van doors were replaced with an innovative swing-plug single piece sliding door which sits flush with the bodyside when closed.

One wall of the luggage van was to house emergency equipment including two types of drawbar which would be used in the event that the train needed recovery by a conventional locomotive or another HST.

Marketing of the new train was to be an important element; the production train was to be known as the High Speed Train (HST) and BR was keen to highlight the speed element of the new train so it was also christened 'Inter-City 125'. This legend was carried on every power car and featured heavily on all advertising and within timetables. Passengers could soon determine if they would be travelling on the new train by locating the '125' symbol in timetables. They travelled on the IC125 in considerably increased numbers with demand rising by one-third in the first three years. The 125s were a revolution in speed and comfort. On the Western Region they generally replaced 90mph trains, many formed of steam-heated Mk1 carriages. Never before or since had there been such a marked improvement in services in the UK.

LEFT No 43002 was returned to original livery as part of the 40th anniversary celebrations in 2016, and named *Sir Kenneth Grange* after the designer of the iconic nose shape. The first production power car is seen running at speed through Coalpit Heath on 10 July 2017. *(Chris Hopkins)*

Chapter Two

Anatomy of a power car

A light-weight power car positioned at each end of the Inter-City 125 delivered the horsepower required to swiftly accelerate the train to 125mph. This chapter takes a detailed look at the components which make up these locomotives. Dual sourcing of many items resulted in various detail differences between batches as built.

Note: *A number of illustrations appearing in this chapter are drawn from documentation produced around the time of Inter-City 125 introduction. For brevity the individually numbered items on these illustrations are not detailed, but key components and their function are explained within the text.*

OPPOSITE No 43118 *Charles Wesley* leads the 12.00 London King's Cross to Inverness 'Highland Chieftain' service through the rugged Scottish countryside near Dalwhinnie on 26 May 1993. By now, the popular Inter-City 'swallow' livery had been applied to almost all IC125 power cars and sets. The daily daytime service from London to Inverness has been a feature of the timetable since May 1984. *(Bill Watson)*

A total of 197 power cars were built at Crewe Works with the intention of forming 95 complete Inter-City 125 trains while providing seven spare power cars. These were for replacement of defective vehicles removed from their sets for unplanned repairs. Four main batches of power cars were ordered with a smaller 'top-up' batch towards the end of the production run.

The first batch (Lot number 30876, ordered on 22 January 1974) formed 27 sets for the Western Region to operate trains from London Paddington to Bristol and South Wales, and included power car Nos 43002–43055. The second batch (Lot number 30895, ordered 24 December 1974), formed 32 sets for the Eastern and Scottish Regions to operate trains from London King's Cross to West Yorkshire, the North East and Edinburgh and included power cars Nos 43056–43119. This order also included four spare power cars, No 43120 and No 43121 for the Western Region, and No 43122 and No 43123 for the Eastern Region.

The third batch (Lot number 30941, ordered 4 April 1978) formed 14 sets for the Western Region to operate trains from London Paddington to Devon and Cornwall and included power cars Nos 43124–43152. It was intended No 43124 would be another spare, but eventually, just 13 sets for that route were formed and No 43151 and No 43152 were also designated as spares.

The fourth batch, (Lot number 30946, ordered 1 December 1978), was intended to provide 18 sets for North East to South West inter-regional services, but priorities had changed and upon delivery five sets with power cars Nos 43153–43162 were added to the Eastern Region fleet. The remaining 14 sets with power cars Nos 43163–43190 were allocated to the Western Region to operate inter-regional services.

A further small batch (Lot number 30968, ordered 24 June 1980), included just four sets but by the time these were delivered, deployment of the fleet was being examined yet further with the goal of maximising the earning-potential of the valuable asset against a background of economic decline. The first set with power cars Nos 43191 and 43192 was added to the North East to South West route fleet and the remaining three sets, with power cars Nos 43193–43198 were delivered to the Western Region. Almost immediately, they were reallocated to the Eastern Region fleet, enabling them to provide sets to begin IC125 operation on the line from London St Pancras to Nottingham and Sheffield. Although the orders were spread over six and a half years, the deliveries ran continuously in a steady stream from February 1976 until August 1982.

For all planned maintenance it was anticipated the trains would stay in their fixed formations and whilst it was found to be relatively feasible for the trailer sets to be kept

RIGHT No 43003 waits to depart from Reading while working the 11.35 Swansea to Paddington on 11 October 1986. This illustrates the original guard's accommodation in the noisy, uncomfortable area at the rear of the power car, and its replacement in the end of a standard class coach designated TGS. This gave the guard the creature comforts taken for granted by the passengers. *(Kevin Daniel)*

together, the additional maintenance needs of the power cars made keeping them paired impossible. On the Eastern Region, the notion of keeping the power cars together with their intended trailer sets was abandoned even before all their first batch had been delivered in early 1979. The Western Region attempted to keep the formations together with mixed success until abandoning the pairings by around 1983.

As far as classification was concerned the power cars were regarded as vehicles within a multiple unit. Production sets were designated Class 253 (for Western Region-based seven-car sets) and Class 254 (for Eastern and Scottish Region-based eight-car sets). Power cars were designated Driving Motor Brake (DMB) for power cars up to No 43152 and Driving Motor (DM) for power cars numbered 43153 onwards. The change in classification came about following the relocation of the guard's accommodation from the rear of the power car

LONGEST SINGLE JOURNEY

The longest individual journey by IC125 was Virgin Cross-Country's 1V61 08.55 Aberdeen to Penzance service in 2003 – a total of 723 miles (and nearly 13 hours) on one train, if you travelled on it throughout, via Edinburgh, Carlisle, Birmingham and Bristol.

into a far more comfortable section at the end of a passenger coach (of which more later).

In 1988, the power cars were re-designated as Class 43 locomotives although this was simply a paperwork exercise. The power cars could still only work in tandem with each other at either end of a rake of IC125-configured Mk3 trailers.

Before looking at the internal layout of a power car in detail it is worth highlighting that inside these vehicles there are five distinct, separated areas. These are the driving cab (a single driving cab on all production IC125 power cars), then the clean air compartment which houses the alternator and all the electrical control equipment, then the engine compartment followed by the radiator compartment, and finally, the luggage van/guard's area.

When sitting in the driver's seat the 'A' side is on the left, the 'B' side is on the right, which originally referred to the engine bank labelling. All British locomotives have a No 1 end and a No 2 end, as the power car has a single cab this is the No 1 end. The pictures showing the inside of the power car were taken in the early 2000s, and although some minor modifications had taken place during the first 25 years of use, the power car still included all the major equipment as installed from new.

This chapter and the descriptions provided relate to the power cars in their 'as-built' condition. Significant equipment changes (i.e.

ABOVE The low evening sun casts long shadows as the 14.52 Aberdeen to London King's Cross sweeps through Stonebridge (just south of Durham), with No 43106 at the head of the train on 22 June 1988. Within three years, electrification of this line would see most trains switch to electric traction allowing the IC125s to be cascaded to other routes. However, the through trains from Aberdeen to London were to remain in the hands of the type until 2019. (Bill Watson)

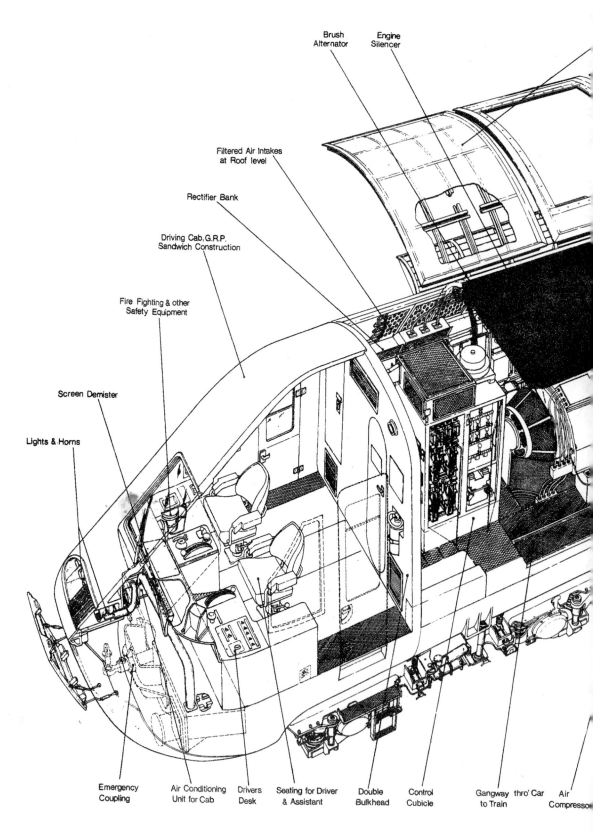

Brush
Alternator

Engine
Silencer

Filtered Air Intakes
at Roof level

Rectifier Bank

Driving Cab, G.R.P.
Sandwich Construction

Fire Fighting & other
Safety Equipment

Screen Demister

Lights & Horns

Emergency
Coupling

Air Conditioning
Unit for Cab

Drivers
Desk

Seating for Driver
& Assistant

Double
Bulkhead

Control
Cubicle

Gangway thro' Car
to Train

Air
Compressor

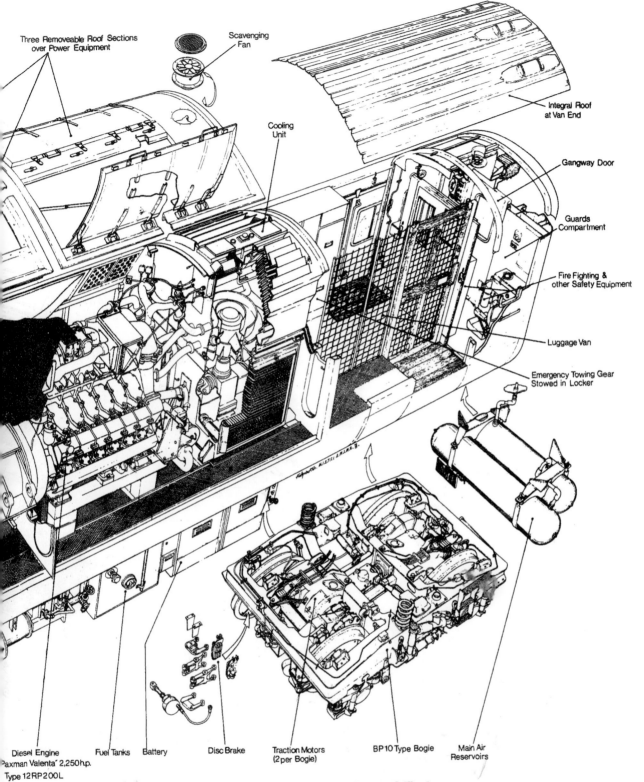

Three Removeable Roof Sections over Power Equipment

Scavenging Fan

Cooling Unit

Integral Roof at Van End

Gangway Door

Guards Compartment

Fire Fighting & other Safety Equipment

Luggage Van

Emergency Towing Gear Stowed in Locker

Diesel Engine 'Paxman Valenta' 2,250 h.p. Type 12 RP 200 L

Fuel Tanks

Battery

Disc Brake

Traction Motors (2 per Bogie)

BP 10 Type Bogie

Main Air Reservoirs

Arrangement of Power Car for High Speed Train

Showing the layout of Power Equipment

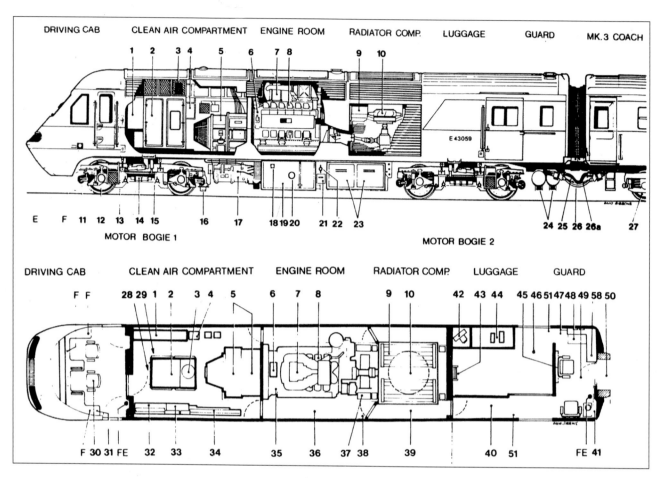

ABOVET The basic layout of a power car, with the key compartments labelled.

those where substantial equipment has been replaced rather than added) are detailed in Chapter Six. A note about units of measurement: for brevity, where an item has a labelled gauge the units are only shown as they are on that piece of equipment, so imperial measurements are used for distances (*e.g.* 125mph), and metric measurements are for air pressures (*e.g.* 7bar).

Technical specifications on liquids and air pressures on power car equipment were generally set in imperial measurements, but the equivalent metric quantity is shown for clarity. The two most important values when operating an Inter-City 125 translate thus: the maximum speed of 125mph is equal to 201kph, and 5.1bar to release the brake is equal to 74psi.

The power car underframe and stressed skin bodyside is of integral all-welded, mild steel construction. The underframe is constructed using 10mm thick material and the bodyside skin is 2mm thick. There are two transverse bulkheads providing additional strength to the structure. These are located between the driving cab and clean air compartment and between

the cooler group compartment and the luggage van. The floor area between these bulkheads has a self-draining sealing plate and spillages are collected in a large flat tank below the engine and alternator. Adjacent to the clean air and engine compartments are air intakes mounted at the top of the bodywork and within these are dry disposable air filters. Four lifting/jacking points are provided and these are concealed behind covers located to the rear of the cab and ahead of the luggage van door on both sides.

Driving cab

The IC125 cab is formed of a Glass Reinforced Plastic (GRP) moulding and is a monocoque construction using a sandwich of GRP material giving it the required strength. The cab sits on top of two steel beams running the length of the power car and is attached to the steel bulkhead using a series of large bolts. Its construction enables it to be removed and replaced allowing for maintenance and accident damage repairs.

A large windscreen provides a good view to the front of the train as well as providing adequate protection to the driver. The 25mm thick laminated glass is resistant to a high level of impact. Use of GRP to produce such a large item was revolutionary at the time of construction. The principal advantage came from the relatively cheap and simple mass-production of the driving cab from a single moulding to a consistent shape resulting in a better fit for glass and internal parts. A disadvantage was a lack of impact strength when compared with steel-framed driving cabs and the difficulty in repairing minor damage without replacing the whole cab. Cabling runs through trunking located underneath the floor, emerging in the control cubicle located in the clean air compartment.

ABOVE The cab for No 43140 on the workshop floor at Brush awaiting refitting. The holes for the bolts which will secure the cab to the bulkhead can be seen through the doorway.

BELOW The cab for No 43149 points upwards, showing the sound-proofed floor underneath.

BOTTOM Lifting point following corrosion repairs, also showing the gap between cab and bulkhead has been filled prior to painting.

BELOW The front structure before the cab has been refitted, showing the bulkhead to which it will be mounted, and the non-operational buffers normally hidden by the cab. The yellow handrail is a temporary addition while engineers work inside the power car.

1 **AWS (Automatic Warning System) – warning sunflower.** Standard equipment on all British trains, the sunflower goes completely black whenever any AWS indication is received. When a warning horn (*i.e.* approaching any adverse signal or speed restriction) is received the AWS reset button must be pressed and the AWS indicator will then go from black to black/yellow to act as a visual reminder of the warning being cancelled. Works in conjunction with the AWS sounder detailed later.

2 **DRA (Driver's Reminder Appliance).** This is a retro-fitted item fitted to all passenger trains where 'right away' (starting signal) is normally given by a buzzer code. The set features a round button which, when pressed, illuminates red and prevents traction power from being taken. The driver sets the DRA by pressing the button whenever arriving at a red signal or stopping when a cautionary signal was displayed at the last signal.

The DRA should be released when the signal changes and when stopped between signals release of the DRA should act as a reminder to the driver of the previous restrictive signal. Next to the button on the DRA set is an isolation switch which is protected by a glass seal, and isolation of the DRA disables the DRA equipment in the event of failure allowing traction power to be taken with the button set or illuminated.

3 **Lighting Supply Socket.** This is one of several 110V sockets positioned around the power car which allow maintenance staff to plug in inspection lights. Power supply is taken from the power car's batteries.

4 **Main Reservoir Pipe Duplex Gauge.** This indicates the air pressure in both No 1 and No 2 air reservoir tanks (left gauge) as well as the main reservoir pipe (right gauge). The gauge is scaled 0–14bar. The left-hand

gauge should normally show between 8.48 and 10bar. The right-hand gauge should normally show around 7bar. If the engine is shut down, or the compressor is defective in that power car, both gauges will read 7bar.

5 **Brake Cylinder Gauges (Power Car).** This shows the pressure in the brake actuating cylinders on the local power car's bogies. The left gauge indicates the front bogie and the right is the rear bogie. The gauge is scaled 0-7bar where 0 is the brake fully released.

6 **Brake Pipe Pressure Gauge.** This indicates the pressure in the train brake pipe and is scaled between 0 and 6bar. The gauge shows 5.1bar when the brake is fully released, 3.35bar when a full service brake is applied, and zero when the emergency brake is applied.

7 **Driver's clipboard** with lights both sides.

8 **Brake Test Switch.** This has two positions: 'Normal' and 'Test'. In the Test position the parking brake protection system is isolated allowing the brake pipe to be charged with air. If the parking brake is applied with the switch in Normal the brake pipe vents to atmosphere. This switch is moved to Test during a brake continuity test as it allows the train to be held stationary using only the parking brake with the train brake fully released. When in the Test position, traction power is isolated so this switch can be used if the train is low on air. By opening the power handle the engine revs will increase, speeding up the compressor, but no power will be applied to the traction motors.

9 **Windscreen Wiper Control.** Turns left for slow, right for fast, centre for off. The button in the middle sends fluid from the washer bottle on to the screen and the whole system is air operated. The wiper motor is accessed via a cover on top of the dashboard.

10 Top left to right: ETS Off Button, ETS Indicator Light, ETS On Button, Brake Overcharge Button. Bottom left to right: Parking Brake Off Button, Parking Brake Indicator, Parking Brake On Button, Fire Alarm Test Button. The Electric Train Supply (ETS) system provides three-phase power from the auxiliary alternator to the train and is detailed later in this chapter.

Operation of the Brake Overcharge Button allowed the train brake to be overcharged to approximately 5.4bar to improve brake distributor functionality, and to overcome a dragging brake. These functions are redundant on a fixed-formation train like an IC125 and so this button was isolated early in the lives of the fleet.

The parking brake operates using a sprung tread brake held off by air. The Dowty-type flag indicator either reads 'Off', 'On' or displays a chevron pattern depending on the state of the parking brakes. Use of a flag indication allows the indication to be seen even if the power car is electrically isolated. The Fire Alarm Test Button not only tests the fire alarm bell and control systems but also locally stops the power car engine, which is useful at terminal stations as it leaves the other engine running and supplying ETS to feed train auxiliaries.

11 **AWS (Automatic Warning System) indicator.** Normally blue but shows red when the cab is off (key out) or the AWS is isolated or defective. AWS consists of track-mounted magnets ahead of signals which indicate clear or cautionary/danger signals to the driver.

12 **General Fault Light.** Normally blue but will go red should a general fault on either power car be detected.
General fault shows if power is being applied in any notch if:
- Main reservoir pressure falls below 4.2bar
- Rectifier Fault Relay (RFR4) has tripped
- Brake pipe pressure falls below 3.0bar
- Power-Earth fault (PEF) has been detected
- High Water Temperature trip has been activated
- Main alternator has overheated
- Rectifier has overheated
- Either engine stops

General fault shows if the power controller is moved to notch one from off if:
- Main reservoir pressure falls below 4.2bar and has not risen above 5.72bar
- Brake pipe pressure falls below 3.0bar and has not risen above 3.9bar
- A serious power earth fault exists on either power car

- Brake Test Switch is still in Test
- Notch 1 selected immediately after shutting off power
- Power car has been isolated

13 **Headlight Control.** Four positions, which are 'Off', 'Left', 'Right' and 'Both'. Switch must be set to Off in a trailing power car. Normal operation was to use the Right headlight during the daytime to avoid dazzling track-side staff, and Left at night to illuminate lineside signage. Some companies changed policies so that Both should be selected partly for aesthetic reasons and partly to ensure there would still be one illuminated if either failed. As built, the headlights draw power from the train ETS supply and this must therefore be in operation for the lights to work.

14 **Windscreen De-mister and ETS Indicator Dimmer Switches.** The windscreen contains an electric heating element. The ETS indicator light can be dimmed to avoid distraction while driving in darkness.

15 **Switches** for front marker lights, front red tail lights (two switches, control lights individually), desk light switch (for the clipboard), engine compartment light switch, and cab light switch. This switch is repeated by both doors.

16 **Emergency Brake Plunger.** Striking the plunger vents the train brake pipe to atmosphere (zero) causing an emergency brake application. This does not require any electrical involvement and is released by pulling the plunger up.

17 **Brake Controller.** This handle has eight steps from Step 0 (running – brake released) to Step 7 (emergency). Steps 0 to 6 are controlled by the electrical brake controller (E70 or DW2, see later) and Step 7 has the same effect as striking the plunger described above. In each step except emergency a reed relay is activated by a magnet on the brake controller which indicates to the control system which step is selected.

18 **Washer Bottle Filler.** Washer fluid is pushed from the bottle to the wiper washer nozzles using compressed air. Below the filler is a removable panel, behind this are air isolating cocks for the horns and wipers, and access to the washer bottle itself.

19 **Driver's Safety Device (DSD) pedal.** Pressure must be applied to this pedal whenever the master switch is in 'forward' or 'reverse' otherwise a brake application will be made. Since construction the power cars have been modified so this pedal now also includes a vigilance function. Every 60 seconds or so a beeper will sound and the driver must release and reapply pressure to the pedal to avoid an automatic brake application. The timing between warnings was controlled by an air valve so it varied between power cars. Operation of the AWS warning reset the timer on the original installation, but subsequently an electronic system was installed which was far more consistent and enabled the timer to be reset by the operation of other controls. Above the DSD pedal is a removable cover which houses the AWS bell, DSD bleeper and horn solenoids.

20 **Speedometer.** On all power cars as built the speedo has a maximum indicated speed of 150mph with a mark at 125mph to show the maximum design speed of the train.

21 **Ammeter.** This shows the power output of the power car and the dial is calibrated in amperes (or amps, for short) and indicates up to 2,000amps full scale. The ammeter monitors the current consumed by the leading traction motors and is simply scaled up by a factor of 2 (as the traction motors are wired in series/parallel). The current is measured using a transformer which is a common means of measuring high currents. When pulling away an average power car will show around 1,650amps, when in notch 5 at around 30mph, which will then drop to around 1,000amps at 70mph (both speedo and ammeter pointing straight up), falling away to around 600amps at full speed in maximum power.

22 **Clock.** When new the power cars had an analogue clock here adjusted by Allen keys. This was replaced by a digital clock although this generally fell into disuse/disrepair and was either left in position or removed altogether and the position blanked over.

23 Guard/Driver Telephone. This is a push-to-talk handset with a button on the handset which must be pressed in order to be heard by the other person.

24 Wheelslip Indicator. Normally blue but will flicker red if wheelslip is being experienced on either power car. If the wheelslip light remains red, even after closing and reopening the power handle, this indicates a traction motor fault and a rotation test (visually checking the wheels are freely rotating) should be carried out as soon as possible. This lamp will also illuminate red and power will be lost if train speed exceeds 132mph. This was a retrospective change as there was no limitation on train speed as built, and as such, 140mph plus runs were recorded in normal service.

25 Engine Stopped. Normally blue but will show red if either power car engine has stopped.

26 Engines Stop and Start Buttons. Operate engines in both power cars.

27 Driver/Guard Buzzer. The function of this button is two-fold: it provides a means for the guard to give the driver the starting signal when leaving a station, and for the driver to confirm this to the guard. It also provides a means of raising the attention of either train-crew that the other wishes to speak through the telephone handset. The buzzer codes are standard throughout the UK railway network with most commonly used in day-to-day operations being one beep indicates stop, two beeps indicates 'ready to start', while the code three-three indicates that driver or guard wishes to speak with the other party.

28 Power Controller. The controller has six positions, 'Notch-off' and five power notches from one to five. In each notch a switch is operated by a cam on the power controller shaft which indicates to the control system which notch is selected. When the controller is moved from Notch-off to Notch 1 the traction motor contactors close, connecting the rectifier bank to the traction motors. The load regulator will then start to control the power from the main alternator by increasing the excitation

BELOW Cab desk, secondman's side.

current in its field. As power is increased or decreased alternator excitation is increased or reduced which changes traction current and power output accordingly.

29 **Master Switch.** This control 'enables' the use of the power car cab controls and is colloquially known as 'opening the desk'. The control has four positions: 'Off', 'Reverse', 'Engine Only' and 'Forward'. On entering the cab the driver puts their key in the hole in the middle of the handle and turns it. They then move the handle to Engine Only. This will cause the AWS warning indication to sound (which is cancelled by pressing the AWS reset button) and since installation during 2000/01, this also initiates the TPWS (Train Protection and Warning System) self-test.

Once in Engine Only all controls are operational except the power controller which is locked from being used. When moving the switch to Forward or Reverse the DSD (Driver's Safety Device) will become active and will sound unless the associated pedal is depressed. Once in Forward or Reverse the power controller becomes 'unlocked' and power can be applied.

30 **AWS/TPWS Reset.** This button cancels any AWS warning that occurs and since installation of TPWS it resets a TPWS brake demand. The AWS warning is given via the 'sunflower' (item 1) and an audible horn. TPWS indications are given by a three-lamp panel which was originally located under the NRN radio to the left of the driver.

31 **Horn Control.** Four positions: 'High Loud', 'Low Loud', 'High Soft' and 'Low Soft'. The control operates four solenoid valves behind a panel in front of the driver's feet which allow air passage into the horns. The

soft positions are used in depot and station areas where the train is moving at slow speed. The loud positions will generally only be used at speed in open country.

32 **Cab Air Conditioning Controls.** These are located on top of the new cab air-conditioning unit installed here in place of the unit under the cab nosecone. The original system had a four-position switch located under the washer/wiper (item 9) enabling 'High heat', 'Low heat', 'Cooling' (which turns the air-conditioning on) and 'Vent', which simply draws fresh air in from under the nose of the power car. There was deliberately no 'Off' position to ensure that air is always circulated in the cab.

33 **Fire Bottle Pressure Light.** Only fitted on INERGEN-fitted power cars. This indicates the fire bottles are still fully pressurised and the fire system is working normally.

34 **Fire Pull Handle/Button.** The original fire system comprised an 18.2kg bottle of Bromochlorodifluoromethane (BCF) which unsurprisingly was usually referred to by the trade name of Halon. This bottle was located under the cab desk on this side of the cab but it was replaced from 2000 onwards by a new environmentally acceptable INERGEN (inert gas extinguisher) system using far larger bottles located in the luggage van (see later). As built, the power car fire system was operated using a handle, but when the system was changed the handle was replaced with a button.

On both systems the fire control is located behind glass and operation of the handle/button will cause the fire extinguishers to operate on the local power car. Also on both systems there is another external fire pull handle located on either side of the cab to allow the system to be activated from track level. The leading power car fire extinguishers will not operate unless the fire pull handle/pushbutton is operated and the master switch is turned to Off. The fire extinguishers activate automatically on the remote power car after 25 seconds with the original system, and once the train speed falls below 6kph on INERGEN-fitted power cars. The need for the unattended rear power car to self-activate its own fire system was a mandatory requirement.

35 Fire Bottle Inspection Flap. As fitted from new until INERGEN installation, a panel lifts up on the top of the desk on the secondman's side and underneath this cover is the fire bottle. A nipple on top of the bottle will protrude if the system has been activated. A transporter bolt is fitted and if the system is normal the tapered end of the bolt is visible. The top of the fire bottle has an electrical connection attached.

36 DSD isolation. This isolates the DSD system on the later electronic system.

Located out of view on the secondman's side near the door is the Emergency Equipment Cupboard which contains two track circuit operating clips. When placed across the track they revert signals to danger in track-circuit-controlled areas. There is also a red flag and a sealed box of ten emergency detonators, which are small explosive charges placed on the running line to warn drivers of following trains there is danger ahead.

Clean air compartment

37 Low Water Light. Normally blue but will show red if the coolant level in the header tank drops to a level where it will shut down the engine and not allow it to be restarted.

38 Control Cut-Out Switch. Two positions: 'Normal' and 'Off'. The latter isolates the

LEFT On-Train Monitoring and Recording (OTMR) was retrofitted to all power car cabs. This is the data entry panel for an Arrowvale-type system.

LEFT OTMR data entry panel for a Q-Tron-type system.

LEFT The roof sections over the clean air and engine compartments contain translucent panels to provide additional light within these areas. The forward roof sections are demountable and those above the engine and alternator are hinged allowing access to the equipment from above.

LEFT The front face of the electrical cubicle showing controls, MCBs and lamps described in the text.

traction motors from the control system and alternator allowing the power car to continue running only to provide Electric Train Supply (ETS) for the train. This switch would commonly be used following overheating or a power-earth fault and should also be used on a totally dead power car as operation of the switch extinguishes the associated 'General Fault' and 'Engine Stopped' lights. A power car cannot be restarted with the switch set to the Off position.

Located out of view above this switch is the Fire Bell Isolating Switch which has two positions: 'Normal' and 'Isolate'. The latter isolates the fire bell and fire detection system and when isolated the Control Cut-Out Switch must also be operated so the power car cannot provide traction power. The fire alarm will still sound on both power cars should there be a further outbreak of fire on the remaining operative power car.

39. Power Cut-Out Switch. Two positions: 'Normal' and 'Off'. This switch is for use only by maintenance staff and is similar in operation to the Control Cut-Out Switch except the engine speed may be varied without the alternator providing any power. The fire detection system, fire bells and power control circuits are inoperative when this switch is set to Isolate.

40. Battery Charge Ammeter. This meter shows the charge or discharge from the power car batteries and is calibrated in amperes. The mid position indicates there is no net discharge or charge, full scale to the left indicates maximum charge (an indicated 150amps) and full scale to the right indicates maximum discharge (an indicated 150amps). Charge will only be provided if the ETS is switched on.

The ammeter will not indicate Charge if the Battery Circuit Breaker (see BCB on MCB panel) has tripped. A small difference in the wiring on power cars Nos 43002–43152 as built means that if the Battery Isolator Switch (external) is open and ETS is sent to that power car from the other power car or from a shore supply, the Battery Circuit Breaker may trip. (No 43121 was later modified to trial the revised wiring arrangement for a period.)

The original wiring configuration on those power cars allowed the engine starter motor to operate with the BIS open which caused safety concerns, so wiring was retrospectively modified. The needle must register on the Charge side of the scale after the ETS is turned on. If the battery is not charging then, after a period of time, the power car will become unserviceable due to a drop in supply voltage to the various control and brake systems. If this occurs driving from the other end or rescue are the only options.

41. Auxiliary Earth Fault Indicator. These three red neon lights are repeated on panels at the end of each carriage. Each light is connected to one of the phases of the ETS system and if one of the lights is not showing it indicates a possible earth fault on that phase of the train supply; the train can continue normally on two phases. Drivers will often check these three lights and the Battery Charge Ammeter after starting a power car and switching ETS onto that power car in order to check all is well with both items.

42. MCB (Miniature Circuit Breaker) Panel. Each circuit breaker is 'off' in the UP position and 'on' in the Down position. Where circuit breakers are linked together they are protecting equipment on the 415V three-phase side of the power car electrical system. The single-circuit breakers protect 110V DC items.

On a typical power car the circuit breakers are arranged as follows:
Top row (left to right)
- Battery Charge Circuit Breaker (BCB). If the BCB trips the battery charge will be lost, but this will not be apparent unless the battery charge ammeter (see item 40) is observed.
- Heater Circuit Breaker (HCB1). Protects power to the windscreen de-icing system.
- Heater Circuit Breaker (HCB2). Protects the cab air conditioning system.
- Heater Circuit Breakers (HCB3+4+5). These HCB's were for the guard's compartment and as such, became redundant, so these positions have been used for other retro-fitted items.

- Control Circuit Breaker (CCB). If the CCB trips this is known as the 'graveyard syndrome', the effect being everything goes very quiet! If this MCB trips in either power car, the train will come to an emergency stop and cannot be moved until it has been reset.
- Train Supply Circuit Breaker (TSCB). This protects the control circuit for the train supply contactor and the interlock wire that loops around the train.
- Public Address Circuit Breaker (PACB). The PACB is ambiguously labelled as it cuts out the driver-guard communication (phone and buzzer) and it is occasionally deliberately tripped if the buzzer is sounding continuously or randomly in the cab.
- Automatic Warning System Circuit Breaker (AWSCB). The Automatic Warning System (AWS) will sound continuously and an emergency brake application will be made if this trips. If reset is not possible then the AWS would have to be isolated to continue. Since installation of Train Protection and Warning System (TPWS) this MCB has been renamed TPWSCB, but has the same function as TPWS is an 'add-on' system that works in conjunction with AWS.
- Fuel Lift Pump Circuit Breaker (FPCB). This MCB protects the fuel lift pump and if tripped the engine will stop due to fuel starvation.
- Lighting Circuit Breaker (LCB1). Protects the driving desk lighting, both marker lights and the left-hand tail light. It also protects some of the lighting around the clean air, engine, radiator and luggage

van compartments, as well as four of the inspection power points.
- Lighting Circuit Breaker (LCB2). Protects the driving compartment lighting, right-hand tail light and the remainder of lighting and power points around the compartments. There are two LCB circuits (one on each side) allowing the power car to continue in service if one alone trips.
- Parking Brake Circuit Breaker (PBCB). Protects the parking brake system. If this trips the parking brake indicator (see item 10) will show the chevron pattern and the Parking Brake Isolation Switch will not operate.

Bottom row (left to right)
- Fire Control Circuit Breaker (FCB). When tripped the fire alarms will ring continuously and the fire extinguisher equipment will be inoperative. The engine of the affected power car will also stop and if tripped and unable to reset the control cut-out switch (see item 38) will have to be operated.
- Brake Control Circuit Breaker (BCCB). If this MCB trips in either end the train will come to an emergency stop. If it cannot be reset, the brake pipe pressure control unit (see later in this section) will need to be isolated and the train can only then be driven from the other end, or rescued.
- Lubricating Oil Pump Circuit Breaker (LPCB). Protects the lubricating oil pump. The starter motor will not crank the engine if this MCB has tripped as the oil pressure will fail to reach the required pressure.

ABOVE No 43022 sweeps round the curve at Briton Ferry on 15 April 1991, at the head of the 17.32 Swansea to London Paddington. As this view shows perfectly, the line through South Wales is quite different from the far straighter 125mph race-track found on the Bristol to London section. Also, freight flows such as those using the adjacent yard, share the tracks used by IC125 services. *(Martin Loader)*

Fuse Box. Each fuse type is a different physical size so it is impossible to use the wrong fuse in any application. As an aide-memoir staff are taught that the fuses standing upright are 'working' *i.e.* they are doing something while the fuses lying flat are 'lazy' *i.e.* spare!

43 Battery Charge (BCF1, BCF2, 80amps).
The battery will not charge if either fuse is ruptured. If either of the fuses ruptures, it may also trip the Battery Circuit Breaker giving the effect of a Control Circuit Breaker trip – *i.e.* no control supply to cab. Again, earlier wiring on power cars Nos 43002--43152 as built, means the rupture of either fuse also stops the lubricating oil pump and fuel lift pump from operating.

44 and 45 Lighting (LF1 and LF2, 35amps).
Protects various lights around the power car (engine room, marker, etc.)

46 Alternator Neutral Fuse (ANF, 50amps).
This is the fuse on the neutral side of the auxiliary alternator and if blown will cause the engine to stop as a result of the Low Voltage Fault Relay tripping.

47 Headlight fuses (HLF1, HLF2, 6amps) which protected the headlight supply from the ETS. They are now repurposed as engine control fuses (ECF1, ECF2, 6amps) and protect the control system for the MTU engine (see Chapter Six) as the headlights are now powered off an MCB.

48 Fuse Tester. This is a simple lamp and contact arrangement that allows for the testing of fuses without the need for a multi meter. To change any fuse or use the tester the ETS should be switched off, the power car engine shut down, and the Battery Isolator Switch (external) operated and the Battery Circuit Breaker should be tripped. The fuse tester is linked to one of the lighting circuit breakers so if that is tripped then nothing will appear to work.

- Blower Motor Circuit Breaker (BMCB1). Protects the front traction motor blowers. If either of the traction motor blower MCBs trip then the cool air that is drawn over the traction motors will cease, power may be lost eventually if they overheat. Additionally, when the MCB for traction motors 1+2 (the front bogie's pair) trips, the cooling for the rectifier will be lost as this pair of traction motor blowers also cool the rectifier.

- Blower Motor Circuit Breaker (BMCB2). Protects the rear traction motor blowers. If it is necessary to open the emergency equipment cupboard in the luggage van then tripping the BMCB2 circuit breaker makes this easier and safer to do as the motor blowers for the rear bogie are housed below that cupboard. The door can become sucked closed or back into place trapping fingers if the blowers are running.

- Compressor Circuit Breaker (CMCB). Protects the compressor motor. If this trips the power car will cease building up air supply, although it is quite feasible for one of the two power cars to build up and maintain sufficient air supply on its own.

- Engine Room Extractor Fan Circuit Breaker (EFCB). Protects the engine room scavenging fan.

On the 'A'-side of the electrical cubicle is a pair of high-tension safety doors and located inside these doors is high-voltage power equipment. If the doors are opened the power car engine will shut down (or not start if already shut down). The equipment located in the left-hand side of the electrical cubical is as follows:

- Left-hand Door. Top (left to right) Fuel Pump Contactor, Lubricating Oil Pump Contactor and Main Alternator Exciter Field Contactor. Each contactor closes to provide power to the relevant machine.
- Left-hand Door. Middle and Bottom (left to right) Four Traction Motor Contactors (two per row) which close when power notches are selected to provide DC power to the traction motors.
- Right-hand Door. Top (left to right) Temperature Monitoring Module (TMM), Alternator Voltage Regulator (AVR) Control Module, AVR Power Module. The TMM measures the temperature of the alternator and rectifier bank. There are two Positive Temperature Coefficient (PTC) thermistors mounted in the alternator and two mounted against the rectifier bank heatsinks. The electrical resistance of the thermistor increases as the temperature increases. The TMM monitors the temperature of all four thermistors and if the temperature of the alternator reaches 190°C or 110°C in the rectifier, the fault relay operates and power from that power car is interrupted. The AVR modules control the excitation of the auxiliary alternator to maintain the variable voltage ETS supply.
- Right-hand Door. Middle (left to right) Load Regulator Demand Module, Load Regulator Control Module, Load Regulator Power Module. These three modules form the combined load regulator which controls both traction power (i.e. load) and tractive effort. The Load Regulator Demand Module monitors the position of the Linear Voltage Differential Transducer (LVDT) which is mounted in the engine governor. This device senses the engine load condition and is used as the primary control signal for controlling the main alternator's field excitation according to the available engine output power.

The engine speed (via the auxiliary voltage) is used to control the traction current in parallel with the governor's traction power control. The control system transitions seamlessly from current to power control as speed rises. This critical main alternator field demand is over-ridden during wheelslip conditions to prevent uncontrolled wheel spin. This overall load demand signal is passed to the Load Regulator Control Module where it is converted into three 'thyristor gating signals' which control the thyristors in the Power Module. The Control Module also has some fault detection circuits which are shown by LEDs mounted on the front of the Control Module and also cause the Fault Relay to trip.

The Power Module controls the main alternator field which is excited using three-phase thyristor rectifiers driven off the three-phase ETS supply. The DC current flow is from the auxiliary alternator neutral, through the ANF fuse, through the main alternator field winding and then through the three-phase bridge. The power module also provides the low voltage power supplies for all three Load Regulator Modules. Power cars with new Brush electronics provide a similar overall result but do not use the ETS to drive the alternators. Supply is derived from the 110V batteries and the exciters are driven using MOSFET transistors.

Below these sits the Wheel Slide Protection (WSP) unit which controls wheel slide (when wheels lock-up during braking under poor railhead conditions). The WSP unit is a 3U rack consisting of 13 plug-in modules. There are two types as the design changed after production of the first batch of power cars so Nos 43002–43055 were fitted with the MkI rack while Nos 43056–43198 had the MkII- type racks. Neither type was interchangeable, but both work in the same manner; each traction motor is fitted with a Brush speed probe.

The WSP unit supplies a 50kHz carrier wave which is modulated by the motor armature ring and processed in order to generate a pulse train corresponding to the relative speed of the traction motor. If the speed of any traction motor drops off when power is removed (i.e. during coasting or braking) relative to the majority, then the brake on that axle is removed by a blow-down dump valve and then reapplied. On the MkI units only the entire bogie brake could be removed and re-applied whereas the MkII units can blow-down any axle. The first batch of power cars were later retrofitted

RIGHT The short circuiter (with yellow and black hazard flash) located on top of the electrical cubicle. To the left of the short circuiter is part of the ATP equipment on a Great Western power car; to isolate the system a driver's key must be inserted and turned.

RIGHT An Arrowvale type data recorder mounted above the electrical cubicle on the 'A' side.

electronics, signals are provided to the electronics rack by the Knorr Bremse WSP system and a two-stage wheelslip correction is provided. If a new slip occurs, power is removed and then reapplied rapidly. If a slip then occurs within a short time frame, power is removed and the power re-applied more slowly.

- Right-hand Door. Bottom Reverser. This is an electro-pneumatically operated off-load switch controlled by the direction of the master control in the cab. The reverser changes the direction of the field windings by using two air pistons and a treadle-type contact bar dependent on the direction selected.

Moving around to the rear of the control cubicle (*i.e.* nearest the alternator) the Short Circuiter, also known as the 'Top hat', is located on top of the control cubicle. Originally this was a Sécheron three-phase type, but later all power cars received a replacement Brush DC type for reliability reasons. This is a very heavy-duty switch which is connected directly to the traction rectifier output and comes into action when the traction output is switched off under fault conditions. Its prime function is to close if a diode failure in the rectifier bank is detected, thus preventing damage to the other diodes, but it also saves costly damage to the motor contactors by closing before excessive power reaches them.

Other items of control and monitoring equipment have been added to the control cubicle here over the years and some of these will be looked at in Chapter Six. Behind the Short Circuiter on top of the control cubicle are the Battery Charge Rectifier and Power Earth Fault Relay Capacitor, the Direct Current to Current Supply Transformer (for the cab desk ammeter drive), and the Wheelslip Supply Transformer.

There are three rows of relays located inside the right-hand side of the cubicle, top row left to right are: Alternator Field Relay (AFR), Brake Control Relay (BCR1), Fault Relay (FR), Low Voltage Relay (LVR), Power Control Relay (PCR), Reset Relay (RSR), Speed Proving Relay (SPR), Train Supply Relay (TSR), Voltage Relay (VR2) and Start Circuit Relay (SCR).

Hidden behind the top row of relays are three Control Rectifier Assembly units (CRK1-CRK3).

with the MkII racks. Power cars with the new Brush electronics have a Westinghouse (later Knorr Bremse) system fitted in place of the BR unit which uses probes on each axle end to provide a reliable speed signal and microprocessor control to account for any number of sliding axle incidents.

Wheelslip (when the train wheels lose grip and turn faster than the train speed when power is being applied), is taken care of as part of the overall control system, together with voltage sensing circuitry that is connected to the traction motors. A passive network of resistors and diodes is arranged across each pair of traction motors such that a 'dead band' is set up where the network is tolerant of voltage imbalances. If the voltage across the traction motors suddenly exceeds this dead band Wheel Slip Relay (WSR) 1 or 2 is activated. The contact closure is then monitored by a simple circuit which is powered by the auxiliary alternator supply, which activates the WSR3 relay if either WSR1 or WSR2 are activated. WSR3 contacts then activate a signal to the load regulator to quickly remove the demand signal and then re-apply power once wheelslip has cleared.

On power cars with the new Brush

On the second row of relays left to right are: Brake Control Relay (BCR2), Stop Relay (STR), Wheel Slip Relay (WSR3), Time Delay Relay (TDR1), Rectifier Failure Relay (RFR4), Power Earth Fault Relay (PEFR2), Over Voltage Relay (OVR) and Time Delay Relay (TDR4).

On the third row of relays left to right are: Engine Run Relay (ERR), Time Delay Relays 2, 3, 7 and 8 (TDR2 etc.), Water Level Relay (WLR), Water Temperature Relay (WTR), Low Voltage Fault Relay (LVFR), Time Delay Relays 5 and 6 (TDR5 etc.) and Wheelslip Control Relay (WCR).

Although each of the relays is vital there are five which drivers learn about as these commonly affect operation of the power car. The Water Temperature Relay (WTR) is a non-latched relay with an indicator flag and is activated by the closure of either one of two Water Temperature Switches (WTS). The relay is connected to the fault relay which, when triggered, will revert the power car to idle via the Power Control Relay (PCR). The general fault light will be illuminated and remain on until the temperature falls sufficiently.

After the engine has cooled down the power controller must then be returned to idle and re-opened in order to regain power from the affected engine. If the WTR trips, the driver has no choice but to manage without power from that engine until it has cooled down sufficiently. This usually takes around four minutes and it can occur during higher ambient temperatures following a period of running in full power (e.g. restarting from a station). It thus places extra demand on the remaining power car which can sometimes cause a further WTR trip, denying any further traction power until one or both power cars have cooled down.

The Rectifier Fault Relay (RFR4) is a latched relay with indicator flag and will trip if the main rectifier becomes defective. This removes traction power but will not cause the engine to stop so the affected power car can still supply ETS. It can be triggered by any of the three RFR relays monitoring the rectifier bank for a fault current. Once RFR4 is latched in, the other contacts enable the Short Circuiter, disable the Short Circuiter from being reset, and open a contact in the engine start circuit. The relay then requires resetting after attention to the fault.

The Power Earth Fault Relay (PEFR2) is a

non-latched relay with indicator flag and is a slave from PEFR1, which monitors the voltage developed at the star-point (centre-point/neutral) of the main alternator with respect to the body/ bogie earth. If the voltage becomes excessive (indicating an earth leakage current of around 5 milliamps is flowing through an insulation fault) PEFR1 and PEFR2 are activated. This causes alternator excitation to be removed and the Short Circuiter to close, resulting in a loss of traction power and a general fault light whenever the power controller is away from notch-off.

If the PEFR2 trips, power can be immediately reapplied, but if the fault persists the driver can try using a lower maximum power demand (i.e. avoiding using notch five) as this will keep DC traction voltage down to a sustainable level. The PEFR circuitry is fairly rudimentary. It will only trip above power notch two and cannot cope with AC earth faults such as those possible from a defective alternator and is then over-sensitive to damp/moisture. The PEFR circuit is four times more sensitive than a household earth leakage while handling voltage five times higher.

The Low Voltage Fault Relay (LVFR) is a non-latched relay with indicator flag which will trip if the auxiliary/train supply voltage falls below a pre-determined value for the current engine speed. It should be possible to restart the engine after an LVFR trip if the fault clears. Although an uncommon fault it is one drivers are taught about, as replacing the Alternator Neutral Fuse can cure the problem allowing the train to resume running. The LVR is part of the Engine Run Valve (ERV) circuit which keeps the engine running and is de-energised in normal operation. During engine start up, LVR is kept out of circuit by the Engine Run Relay and Fuel Pump Contactor. Once the LVR becomes energised it holds the power off the LVFR via TDR7.

The Over-Voltage Relay (OVR) is a latched relay with indicator flag and will cause the engine to shut down and will be unable to be restarted until the latched relay is unset. A transformer, rectifier and low-pass filter provide a DC version of the three-phase auxiliary supply to a comparator circuit. The engine will be stopped by LVFR after eight seconds if the auxiliary voltage is too low, and if the voltage becomes too high in the auxiliary circuit the comparator will activate the OVR.

Eight relays (TDR1–TDR8) control various time-related functions. TDR1 and TDR2 are on the start circuit. TDR1 energises when the oil priming pump reaches 6psi when starting the power car. TDR3 is on the power circuit, while

TDR4 and TDR5 are part of the fire system. TDR4 provides a safety mechanism whereby the fire extinguishers are not triggered by pressing the fire alarm test button, whilst TDR5 introduces a 25sec delay to the fire system activation allowing anyone inside the engine room time to leave before the fire extinguishers activate. TDR6 is activated when the Electric Train Supply (ETS) is switched on and causes contacts to close after 4sec (to allow the engine to reach 1,000rpm), which then holds the Train Supply Relay (TSR) on.

If a fault is detected or the supply is switched off TDR6 immediately drops out which causes TSR to drop out as well. TDR7 is on the auxiliary alternator circuit and energises the LVFR if more than 7sec of low-voltage output is detected. TDR8 is on the brake circuit and prevents a brake release in under 60sec following a brake application initiated by a safety system.

Located below the relays on the left-hand side of the bottom of the cubicle is the Train Supply Contactor (TSC) which connects the three-phase auxiliary alternator output to the ETS cables running throughout the train. The TSC opens if the circuit is broken and it ensures only one power car is providing train supply.

Opposite this side of the electrical cubicle on the 'B' side wall is the main rectifier bank

RIGHT The view down the narrow gangway with rectifier bank behind the cover on right, and the relay side of the cubicle on left. The parking brake isolator switch is above the orange box at the far end.

FAR RIGHT The rectifier bank is seen with the cover removed.

and this houses the three-phase bridge rectifier diodes and heatsink assemblies, which convert the AC current from the alternator into dc current suitable for the traction motors. There were two types of rectifier but both types have the same functionality and outward appearance. Power cars Nos 43002–43152 had rectifiers with 36 diodes whereas Nos 43153–43198 were fitted with the later version of rectifier with 12 diodes.

The diodes are force air-cooled with the air supply for traction motor blower Nos 1+2 which draws air from the louvres on the 'B' bank side of the power car adjacent to the cab door. Also contained within the unit are two temperature sensors for the Temperature Monitoring Module (described earlier) and three Rectifier Fault Relays (RFR1-3) on the incoming three-phase supplies from the alternator. These are designed to trip if there is a major overcurrent from the alternator or rectifier. The rectifiers originally included sacrificial over-voltage protection 'suicide' diodes at the base of the unit, but these were removed, partly for reliability reasons and partly because they were unnecessary.

Located in the leading end corner is the parking brake isolator switch, which has three positions: Normal, Isolated and Isolated. When the switch is in either Isolated position the

THE BASIC PRINCIPAL BEHIND OPERATION OF AN AIR-BRAKED TRAIN

Two of the most important terms used by railway personnel when talking about air brakes are 'creating' and 'destroying' a brake. Creating a brake involves building air pressure up in the brake pipe to release the brake, this means drawing air from the main air reservoir to raise the brake pipe pressure to 5.1bar which is the point at which full brake release is achieved and the train can be moved. The brake is applied by pushing the brake controller forward which reduces the pressure in the brake pipe, a reduction of brake pipe pressure down to 3.35bar gives the maximum possible normal brake application (known as 'full service brake'), while any emergency brake application vents the brake pipe to atmosphere.

The brake is released by pulling the brake controller back. This raises brake pipe pressure; increasing brake pipe pressure back to 5.1bar fully releases the brakes again. On an IC125 this is done in steps with set pressures at each step while on many conventional locomotives the brake is continuously variable. The two systems are entirely compatible, allowing an IC125 to be hauled by a locomotive and vice versa.

Destroying the brake is the process of venting the brake pipe down to zero by the use of the emergency brake prior to removing the driver's key. This empties the brake pipe securing the train and leaves it ready for the brake to be created in either cab. The brake pipe runs the full length of the train and the air pressure in the brake pipe is always controlled using the brake controller in the leading cab. This is a fail-safe system as any disruption to the brake pipe will mean air pressure would be lost resulting in an emergency brake application.

LEFT The last train of the day: No 43117 *Bonnie Prince Charlie* waits at London King's Cross prior to departing with the 23.30 to Leeds on 26 January 2005. No 43106 *Fountains Abbey* was at the front. *(Chris Martin)*

parking brake protection device is energised, which allows the train brakes to be released in the event of air loss from the braking system. The parking brakes must then be manually wound off using the release wheels on the power car bogies. It is worth noting that the parking brake indicator in the cab will still display On, even when the parking brake has manually been released. This switch is a retrofitted item and is only present here on power cars Nos 43002–43152.

Located on the 'A' side wall of the clean air compartment are two brake frames and the brake pipe pressure control unit. The No 1 brake frame is furthest from the cab and the No 2 brake frame is just inside the clean air compartment, close to the cab. These are separated by the brake pipe pressure control unit. The compressor builds up air and supplies this to the main reservoir feed and runs via the brake frame and charges the air reservoir tanks located underneath the rear of the power car.

The main reservoir feed runs at 7bar. A main reservoir pipe runs the length of the train so, in the event of a compressor defect, the other power car can supply air for the whole train. The main reservoir supply on each power car runs at around 10bar when the tanks are fully charged. The key piece of equipment on the

brake frame is the Brake Pipe Pressure Control Unit (BPPCU). This is commonly referred to as the 'E70', but that abbreviation encompasses one of two compatible and interchangeable units which perform the same function. Brake pipe pressure control units can be either the Davies & Metcalfe (D&M) E70 unit (fitted to No 43002–43152 from new), or the Westinghouse DW2 unit (fitted to No 43153–43198 from new).

The brake pipe pressure control units in each power car are linked by three wires which are part of the '36-way' cable running the length of the train. The importance of the E70 cannot be overstated. On a conventional train the brake is applied by the driver at the front and each vehicle in turn reacts to a drop in the pressure in the brake pipe, and the brake is applied on that vehicle. This means the rearmost vehicle in the train will not start applying the brake until several seconds after the driver has started to apply the brake, losing valuable stopping distance. On an IC125 the electrical signals conveyed between each E70 mean the brake is applied instantly at both ends of the train, greatly reducing the amount of time between the driver applying the brake and the last vehicle reacting. The last vehicle to apply is now the one in the middle of the train.

If a defect occurs and it becomes necessary to isolate the E70 on the rear power car, the train is restricted to 100mph as the stopping distance is extended to that of a conventional air-braked train. The E70 was designed as a self-contained item and it comprises electrical control panel, regulator valve, transmission valve, isolating restricting valve and a timing reservoir. As reliability of the E70 is critical, a test facility was set up within Derby Locomotive Works in 1980 enabling it to be overhauled and tested as a complete unit before being refitted to a power car.

Originally, train wires 20+21+22 were used to convey signals for the E70, but wire 22 was on the outside set of cables running through the train wire bundle and was prone to breaking and malfunction. For reliability reasons there was an early change to the wiring to use wire inner cable 15 whose function was swapped with wire 22. A binary signal is used, each wire is either energised or de-energised and a decoder in the E70 interprets the signal via

RIGHT The central section between the two brake frames below the E70.

a comparator which compares the demand voltage with the actual current brake pressure and applies or releases the brake pressure accordingly. If the comparator detects the brake needs applying it will activate the Application Electro-Pneumatic Valve (EPV), which is known as 'valve 49'. This will exhaust air from the control pipe which in turn will exhaust air from the brake pipe via a transmission valve.

If the comparator detects the brake needs releasing it will activate the Release EPV, which is known as 'valve 48'. This will start to charge the control pipe using air drawn from the main reservoir reed pipe. The feed to the Release EPV is via a pressure reduction valve which limits the input to the EPV to 5.1bar. As the control pipe charges it will cause the brake pipe to start to charge via the transmission valve which has a direct connection to the main reservoir supply pipe. Finally, there is a control reservoir on the control pipe to allow smooth brake application and release.

The brake controller (see item 17 on the cab layout description) has seven brake positions (referred to as 'steps') and the E70 creates the signal to send via the train wires to the other E70 as follows:

- 'Running' energises train wires 15, 20 and 21 – brake pipe pressure is 5.1bar.
- 'Initial' energises train wires 15 and 21 – brake pipe pressure is 4.6bar.
- 'Step 2' energises train wires 15 and 20 – brake pipe pressure is 4.35bar.
- 'Step 3' energises train wire 15 – brake pipe pressure is 4.1bar.
- 'Step 4' energises train wires 20 and 21 – brake pipe pressure is 3.85bar.
- 'Step 5' energises train wire 21 – brake pipe pressure is 3.6bar.
- 'Full service' energises train wire 20 – brake pipe pressure is 3.35bar.
- 'Emergency' de-energises all three wires – brake pipe pressure is zero.

The system is fail-safe such that all the train wires must be energised to release the brakes. The coding is also such that any one-wire failure will still allow braking. For example, if any wire becomes permanently energised the driver can still brake the train. Emergency not only de-energises all the train wires but also

activates a mechanical release valve on the brake pipe, venting the pipe to atmosphere, which would result in an emergency brake application, even if all three wires were falsely energised.

A second pair of valves referred to as 'valve 44' and 'valve 100' only allow the front power car to release the brake and prevent the rear power car from creating a brake itself. This means brake release is more gradual than application. On the front power car valve 44 is open, valve 100 is closed and on the rear

ABOVE D&M E70 type Brake Pipe Pressure Control Unit. Note the printed circuit board to the right of the unit has been pulled out of position for this photo.

BELOW Westinghouse DW2 type Brake Pipe Pressure Control Unit.

power car valve 44 is closed, valve 100 is open. Initially, the E70 was fitted with an over-charge facility, and as part of the daily train preparation and following changing ends, the brake system would be overcharged to 5.4bar (78psi) and over a predetermined time the pressure would reduce to the normal working pressure. This was done to make each individual brake distributor work smoothly after a period of inactivity and overcome any imbalance in the brake system between driving cabs.

Whilst this system had worked effectively on previous train types it was found on the IC125 that the combination of slow over-charge release and impatient drivers was resulting in trains being moved before brake pipe pressure had returned to normal throughout the whole train, resulting in damage to wheelsets. An initial modification tried reducing the time taken to release the excess brake pipe pressure, but eventually the overcharge facility was removed completely and replaced by valves on each trailer car that dump the control pressure when the brake is placed in emergency.

The Westinghouse DW2 brake pipe pressure control unit is fitted to some power cars. It performs the same function and uses the same train wires and coding so is entirely compatible with the D&M E70, but internally it works differently being electro-pneumatically controlled rather than electronically controlled. The brake step wires connect to three solenoids on top of a seven-step relay valve. Energising and de-energising these solenoids (A, B and C) controls air into three corresponding chambers. There are three diaphragms with potential pressures of 4, 6 and 7 units. By varying the way in which these are used the corresponding pressures are created so the DW2 output copies the E70.

On the power cars the two bogies have their own brake distributors and supply reservoirs mounted in the brake frame. The distributor takes the brake pipe pressure and drives the required brake cylinder pressure from that. Its function is inverse, so with brake pipe pressure at 5.1bar the bogie brake pressure is zero. As brake pipe pressure reduces the distributor increases brake cylinder pressure. The distributor valves consist of a sprung hollow stemmed valve to apply air pressure to the brake calliper cylinders. When in the running position the brake pipe forces the bottom (sprung) valve closed, which in turn pulls the hollow valve down shutting off the main reservoir supply to the brake cylinder via the top valve. When the brake is applied the reduction in the brake pipe pressure causes the spring to force the hollow stemmed valve up which in turn admits main reservoir air into the brake cylinder via the top valve. When in the lapping (holding) position the brake pipe and trapped brake cylinder pressure equal each other and the brake continues to be applied. When being released the brake pressure increases causing the hollow stem to fall which then exhausts the air from the brake cylinders, thus releasing the brake.

No 1 brake frame
49 Driver's Safety Device (DSD). Electro-Pneumatic Valve (since removed upon installation of electronic DSD).

50 Duplex Check Valve. Closes at 5.0bar.

51 DSD Reservoir. (Former location).

52 Parking Brake Released Pressure Switch. On power cars Nos 43153–43198.

53 Parking Brake Application Pressure Switch. On power cars Nos 43153–43198.

54 Low Main Reservoir Pressure Governor. This is a governor with a toggle switch which prevents brake release with insufficient air pressure in the system, closing at 5.9bar and opening at 4.5bar.

55 Brake Pipe Pressure Governor. This is a governor with a toggle switch which prevents brake release with insufficient air pressure in the system, closing at around 3.9bar and opening at around 2.9bar.

56 Compressor Governor. Coses at 10.0bar, opens at 8.48bar.

57 Compressor Unloader Isolating Cock. The compressor runs continuously and when the air system reaches over 10bar the compressor unloader dumps the excess air from the system via the valve located to the left of the cock. If this becomes stuck open the isolating cock stops the unloader from venting air from the system, and the safety valve will exhaust excess air.

58 AWS Isolation Switch. Operation of this switch prevents the AWS from making a brake application and restrictions apply to the operation of the train if it becomes necessary to operate this.

59 TPWS Isolation Switch. Function as per item 58.

60 FVF2. This is a pressure control valve which reduces the 10bar supplied from the main reservoir supply down to 7bar for the main reservoir pipe.

61 Parking Brake Interlock Electro-Pneumatic Valve.

62 Safety Valve. This is the 'last resort' in the air system and vents the main reservoir if it exceeds 10.7bar. Normally the compressor unloader vents excess air from the system.

63 Compressor Governor Isolating Cock. This isolates the governor from the air supply resulting in the compressor running continuously.

64 Main Reservoir Isolating Cock.

65 DSD Isolating Cock. Operation of this cock prevents the DSD from making a brake application. (Cock and associated exhaust valve since removed upon installation of electronic DSD.)

BELOW Left-hand (No 1) brake frame. The numbered items are detailed in the text.

ABOVE Right-hand (No 2) brake frame. The numbered items are detailed in the text.

No 2 brake frame

66 **No 2 Bogie Air Release Valve.**
67 **No 1 Bogie Air Release Valve.**
68 **Position for Parking Brake Controls and gauge.** On power cars Nos 43153–43198.
69 **No 2 Bogie Electro-Pneumatic High Speed/Low Speed control valve.**

70 **No 1 Bogie Electro-Pneumatic High Speed/Low Speed control valve.**
71 **No 2 Bogie Distributor.**
72 **No 1 Bogie Distributor.**
73 **Brake Pipe Pressure Control Unit Isolating Cock.**
74 **Control and Expansion Air Reservoirs.**
75 **Power Car Bogie Brake Isolation Cocks.** (Located just out of view).
76 **Chokes for the TDAVR.** There are two chokes in this box, AVR Choke 1 and AVR Choke 2. These chokes are used to provide a near constant current supply to the auxiliary alternator exciter winding.
77 **Traction Supply terminal box.**
78 **INERGEN extinguisher nozzle.** Identical nozzles are located elsewhere in clean air and engine compartments.
79 **TPWS system junction box.**
80 **TPWS power supply.**
81 **TPWS control electronics.** Items 79-81 were retrofitted when TPWS was installed.

RIGHT Chokes and other auxiliary equipment located on the 'B' side wall. The numbered items are detailed in the text.

THE BATTERY CHARGER

The Battery Charger is common to the IC125 power cars and Mk3 trailer vehicles. Their function is relatively simple. From a variable frequency, variable voltage three-phase supply it provides a nominal 110V DC supply suitable for charging lead acid batteries on the vehicle. This means dealing with the ETS from 276V/33.3Hz to 415V/50Hz, which is the ETS range when the power car is supplying ETS to the rest of the train. Note the power car which is not supplying ETS, powers its battery charger from the train ETS supply, not its local supply.

The three-phase supply comes through a step-down isolation transformer. The charger then attempts to keep the battery charge within certain pre-set limits, namely up to 35amps output and the voltage is 110V. Then, from 35amps to 75amps, the output current falls at 1V/3amps. At 75amps the output is current limited, which avoids overcharging the batteries and prolongs their

life. The power stage is similar to that driving the main alternator, *i.e.* a three-phase thyristor converter with a large inductor (to smooth the DC output). Fault monitoring is provided for both over-voltage condition, *i.e.* the battery voltage gets to 116V, and also excess battery ripple, *i.e.* more than 40V peak-to-peak AC detected on the battery charge voltage. These faults cause the battery charger circuit breaker to be tripped.

LEFT The battery charger mounted on 'B' side wall.

The alternator converts rotational energy from the engine into electrical energy to supply power for the traction motors, train auxiliaries and electrical power demands on the power car itself. The Brush Electrical Machines BA1001B alternator is a compact, self-ventilating, single bearing alternator set that provides two power outputs from a single unit. Use of an AC alternator on the IC125 marked a significant departure from earlier locomotives which used a DC generator. Frictionless magnetic coupling eliminates brushes and commutator. Diversion of the traction motor fields is not required as the alternator voltage range is high enough to be able to fully utilise the engine power available. This makes the IC125 unique and significantly improves reliability, but does result in a low starting tractive effort. It is not usually a problem with two power cars operational.

The main alternator has a frequency range of 75Hz at idle, to 150Hz at full power and is flange mounted onto the engine with the engine providing the bearing for the driven end of the alternator. A cooling fan is mounted on the flange coupling plate to provide cooling

for the alternator. Air is drawn in through the grilles, along the alternator and then sucked down through a vent in the floor of the power car. The alternator is then split up into six parts. From the engine end these are: Main Alternator, Auxiliary Alternator, Surge Suppression Resistors, Main Exciter, Auxiliary Exciter and Rotating Diode Assembly.

The main alternator is a 12-pole, three-

BELOW The Brush BA1001B alternator in situ.

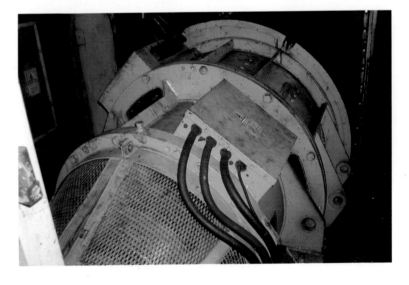

ABOVE The Brush BA1001B alternator viewed from the B side, removed from a power car and detached from the engine.

BELOW The dual main and auxiliary alternator dismantled into its constituent parts with stator, shaft and bearing visible. This is actually an alternator from an Australian XPT power car. Much of the major equipment used on the XPT was the same as used on the IC125.

phase star-connected machine providing up to 1,868kW maximum, but is only allowed up to 1,678kW (2,250hp) in the IC125. The current range is from zero to 1,650A DC with the voltage ranging from zero to 2,300V DC. The star connection provides the three-phase outputs to the main rectifier together with a neutral connection. The alternator consists of a rotating Main Rotor which takes its power from the Main Exciter via the rotating diode assembly. Power to the traction motors is taken from the fixed Main Stator coils mounted on the outside stator frame of the alternator around the rotating rotor. Being a 12-pole machine the alternator has 12 rotor pole bricks on the shaft.

THE ELECTRIC TRAIN SUPPLY (ETS) SYSTEM

Only one power car can provide 415V ETS to the train at any one time. With both power cars functioning normally it is generally the rear power car where the ETS will be switched on. This creates a more comfortable environment for the driver as the engine providing ETS will idle at 1,000rpm rather than 750rpm. It also allows ETS to be conveniently turned on in the leading power car if the supply is interrupted or the rear power car becomes defective during the journey. Pressing the ETS 'On' button in either cab will raise engine speed to 1,000rpm, and if the proving circuit throughout the train is complete, ETS power will be provided from that power car. ETS cannot be drawn from both power cars simultaneously so if already turned on it must be turned off (which can be done from either cab), otherwise attempting to change which power car is providing supply will be unsuccessful.

The full 415V is only available when the engine is running at full speed. The voltage supplied is lower when a lower power notch is used, falling to 276.7V when the power car supplying ETS is idling at 1,000rpm. Equipment throughout the train will still function but there is a discernible drop in the performance of higher demand equipment (such as air-conditioning and kitchen cooking equipment) when the engine speed is low.

The ETS supply is regulated on power cars with original electronics by the Thyristor Divert Automatic Voltage Regulator (TDAVR) whose job it is to keep the output voltage between phases at 8.3x frequency. The TDAVR is actually very simple and monitors the ratio, diverting current away from the alternator exciter when the voltage gets too high, allowing more current through when too low. On power cars with new Brush electronics, MOSFET transistors are used to control the flow of current through the exciter, but essentially the control circuitry is the same.

The traction load regulator generates a DC current in the Main Exciter Stator which causes a larger AC current to be developed in the Main Exciter stator as it rotates. The rotating diode assembly rectifies the three-phase AC before it is passed through the Main Rotor. This DC current causes AC traction power to be generated in the Main Stator windings. Temperature detection probes are fitted to the traction alternator and these cut off power if over-heating is detected.

The auxiliary Brush BAH601B alternator is a four-pole, three-phase star-connected machine capable of providing up to 450kW, 415V at 1,500rpm. This is used for providing power for some of the power car auxiliaries and the Electric Train Supply (ETS) system on the train for carriage heating, air conditioning, kitchen equipment and so forth. It operates in the same manner as the main alternator but this time it is excited by the Auxiliary Exciter. Being a four-pole machine the alternator has four rotor pole bricks on the shaft. The auxiliary alternator works in exactly the same way as the main alternator except that its exciter is controlled by the TDAVR module to provide a controlled three-phase supply at all times.

The surge suppression resistors are mounted on the rotating alternator shaft to prevent damaging voltage spikes on the main and auxiliary rotor coils. Power connections are made to the alternator via two terminal boxes mounted on the alternator; traction power towards the 'B' side and auxiliary power towards the 'A' side. The exciter connections are made to a small junction box on the underside of the alternator towards the free end of the unit.

ABOVE No 43106 *Fountains Abbey* passes Frinkley Lane while leading the 17.20 London King's Cross to Hull GNER service on 18 July 2006. The train will be accelerating back up to maximum speed after having called at Grantham a few minutes earlier. *(Peter R.T. Sharp)*

Engine compartment

Manufacturer	GEC Paxman Diesels, Colchester, England
Model	RP200L (RP = Ruston Paxman, 200 = Bore (mm), L = Locomotive *i.e.* rail use)
Capacity	79 litres
Maximum output	2,250hp (1,678kW)
Number of cylinders	12, arranged in a 60° Vee
Bore	197mm
Stroke	216mm
Aspiration	Intercooled Napier SA-084
Cycle	Four-stroke
Compression ratio	13:1
Direction of rotation	Anti-clockwise, as viewed from the alternator
Idling speed	750rpm
Maximum speed	1,500rpm
Oil capacity	75gal (341 litres)

When the Valenta was introduced the use of four-stroke diesel engines in such applications was normal. Mass-produced large, two-stroke engines without unacceptably high exhaust emissions caused by unburnt oil entering the exhaust were unavailable

ABOVE At the heart of every IC125 power car as built was the Paxman Valenta RP200L engine. The engine was derived from Paxman's Ventura range and although we are only looking at the engine as fitted to the IC125 power car, it can be found in a wide range of other applications including marine and static power generation.

This engine is painted green for a reason. After privatisation the engines were painted according to the train operating company they were 'dedicated' to: green indicated a Great Western engine and behind it is a red-painted GNER engine. The engine is viewed from the 'A' bank side with the alternator attached.

BELOW A Paxman Valenta engine viewed from the free end ('B' bank side) showing the pipework to the adjacent cooler group in place.

until much later. There are four parts to every power cycle of the engine: induction, compression, ignition and exhaust. These are sometimes simplified to: suck, squeeze, bang and blow, as an aide-memoir for how the constantly repeating cycle works.

Induction is where the air is drawn into the cylinder at the start of the cycle, which is essential for the combustion process. The inlet valves are opened while the downward motion of the piston draws air into the cylinder. Compression is where the air is compressed to between 1/13th up to 1/25th of its natural state, causing it to heat up dramatically.

Ignition is when diesel fuel is forced into the cylinder by an injector at the top of each cylinder. This is under huge pressure of up to 2,000psi (140bar). The fuel ignites rapidly, expanding the detonated gas inside the confined space at the top of the cylinder thus forcing the piston down.

Exhaust is where the exhaust valve opens while the piston pushes out the used gas from the confines of the cylinder. The cylinders are numbered from the free end of the engine, so when viewed from the alternator, the left-hand side cylinders are B6>B1 and the right-hand side are A6>A1. The engine firing order is A1, B6, A5, B2, A3, B4, A6, B1, A2, B5, A4, B3.

As built, each engine was fitted with a Napier SA-084 turbocharger. The turbocharger (sometimes referred to as a blower) uses exhaust gases from the diesel engine to drive a turbine which then compresses incoming air for the engine. This allows the combustion process to be more efficient, leading to a higher specific output of the engine. The turbocharger boosts the air flow into the cylinders at pressures of up to 25psi (1.8bar) thus increasing the amount of air volume for combustion.

Fresh air is fed by filtered air vents located on the engine room roof louvered sections. Internally, the duct funnels the air down from the size of the external air vents to a much smaller air intake either side of the turbocharger. The centre of the unit is the compressor/exhaust turbine section and the whole unit is water cooled. All turbochargers use a centrifugal compressor which sucks air in at the middle and forces it outwards into the volute housing. The turbine shaft sits in a housing with both

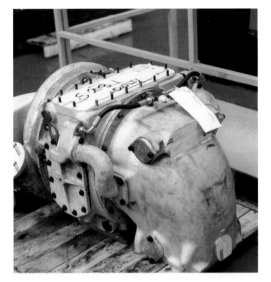

FAR LEFT A Napier SA-084 turbocharger.

LEFT An upside-down view of the SA-084 showing (at the top of the image) the exhaust gas inlet where it would be mounted on top of the engine.

inlet and exhaust chambers sealed from each other. The shaft rotates at up to 27,000rpm and is lubricated from the main engine oil supply.

The volume of air rushing into a turbocharger of this size, at a rate of around 10,000 cubic feet per minute, creates a lot of noise. When running at full speed this drowns out the engine noise and resulted in the distinctive 'scream' produced by the original Valenta-engined IC125 power cars. The air passes through an intercooler which dissipates heat from the compressed air, increasing its density. From 2002 onwards, a replacement NA256 turbocharger was fitted to some power cars in place of the original SA-084 turbocharger. After 25 years of use the original turbochargers were approaching the end of their useable lives and the NA256 delivered considerable operating benefits.

The NA256 performs the same function as the SA-084 but runs slightly slower at 24,000rpm, resulting in a small but discernible difference to the sound produced. The opportunity was taken to improve the way the impeller was manufactured and the coolant flow was slightly simplified. The charge air pressure increased to around 30psi which improved engine reliability, reduced exhaust temperature, and improved fuel economy.

Diesel oil is stored in a pair of tanks located underneath the power car. A continuously running DC motor-driven fuel-transfer pump is located ahead of the No 2 fuel tank on the 'B' side, which draws fuel from a feed at the bottom of the tank and pumps it up to the engine room. The feed from the transfer pump

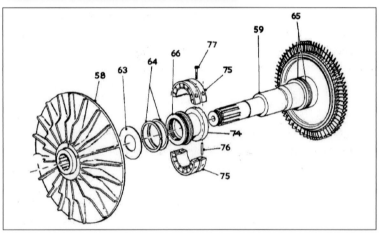

ABOVE An exploded view of the single-shaft turbine as used in the SA-084 turbocharger. It has a turbine wheel at one end driven by the exhaust gases from the engine. At the other end of the shaft is a compressor stage which acts like a large suction fan.

BELOW A NA256 turbocharger.

RIGHT An exhaust silencer – exhaust passes through the two-part water-cooled manifold then through the turbocharger, exhaust outlet elbow, flexible bellows, and finally the silencer before exiting via the two ports.

BELOW A charge air cooler (often referred to as an intercooler or aftercooler) – this takes high-pressure hot air from the turbocharger exit and reduces the temperature of the air by passing it through internal heat exchanger elements so that it increases in density before it is passed to the cylinders.

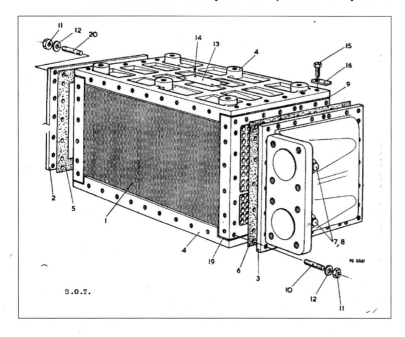

S.O.T.

STARTING THE ENGINES

Starting a Paxman Valenta engine from 'cold' typically takes around 20-30 seconds. First, the lubricating oil pump starts and this raises engine oil pressure and when it reaches 6psi a pressure switch closes. This energises the Governor Run solenoid and the fuel priming pump will start to run. A delay switch means the fuel priming pump must run for at least 8sec. After this a fuel-limiting device on the governor will energise to restrict the amount of fuel injected during starting. The starter motor (a 96V-rated Bosch starter) is then energised, the starter motor dropping out after the engine has reached 175-200rpm, but will only run for a maximum period of 15sec.

When the engine fires, the starter motor and lubricating oil priming pump will be de-energised once engine speed has thrown the centrifugal switch out. When the lubricating oil pressure reaches 25-30psi the red 'engine stopped' lights will extinguish and the engine start button can then be released. If the starter motor has run for 15sec before firing or it fails to engage then the start procedure will be aborted and there will be a delay of 20sec before a further attempt can be made. Stopping the engine is far simpler; pressing the engine stop button will cut the fuel feed pump and de-energise the governor run solenoid so the fuel racks spring back to their 'no fuel' position.

passes through two filtration stages. A main Vokes coalescer filter is located on the 'B' bank side adjacent to the engine and has a five-micron rating. Fuel enters the centre of a cylindrical filter and is drawn from the outer element of the filter material. A coalescer/ agglometer element forms together any water in the fuel into large droplets, which then collect in the sump of the filter.

A Vokes fine filter (often referred to as a secondary or 'chip' filter) is fitted between the coalescer filter and the engine and protects the injection equipment from any debris that may be present in the pipe between the main filter and the engine.

Fuel is forced into the outer element of the filter and drawn from the centre and it has a 10-15 micron rating. The filter is also accompanied by a fuel reservoir header to ensure adequate supply of fuel and pressure. Each cylinder is fitted with a Lucas/Bryce fuel pump. The provision of a pump for each cylinder allows a high output from a comparatively compact engine. The pumps are driven from dedicated camshafts which run along each side of the V of the engine (*i.e.* each camshaft drives six pumps) and can produce a pressure of 1,500psi (100bar). The pumps can be changed individually without having to 'tune' the engine, allowing for simpler maintenance.

ABOVE A fuel pump cam and box.

LEFT A close up view of a fuel pump on a Valenta engine.

BELOW Fuel racks separated from a Valenta.

The fuel injectors are located in each cylinder head and deliver the very high-pressure fuel spray mist down into the combustion chamber. Fuel is delivered to the fuel pumps by one of two fuel rails (one for each side of the engine) and each rail continues after the line of pumps and recombines to form a feed carrying unused fuel back to the No 2 fuel tank.

A fuel priming pump is fitted which feeds fuel up to the engine when starting. This must run for several seconds allowing sufficient fuel to reach the cylinders prior to the engine firing. A local fuel priming pump button is provided on the 'A' bank side near the floor and next to this are local engine Start and Stop buttons.

The lubricating oil system on a Paxman Valenta works in much the same way as any other engine. Oil from the engine's sump is

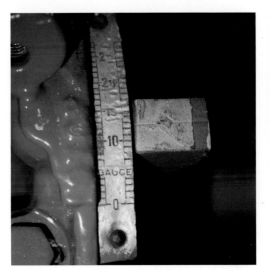

FAR LEFT A fuel rack gauge located at the end of the rack.

LEFT A Bosch starter motor.

pumped using a DC electric oil pump to oil galleries, which then supply oil to all critical parts of the engine including the pistons, crankshaft, camshaft and governor. The turbocharger is also supplied from the engine oil supply. There are two twin-rotor design oil pumps which are mounted in a common casing attached to the engine housing at the drive end directly beneath the crankshaft. The pumps are driven by an idler gear which meshes with the crankshaft gear. There is a pressure relief valve mounted on the free-end cover of the engine and the oil flow that is relieved by the valve is used to cool the external surface of the engine damper. The relief pressure is set at 65psi (4.5bar).

The oil system is of the open-vent design and vents through an external breather located on the roof of the power car on the 'A' bank side. Before the engine can be started a Varley electrical oil priming pump (driven from the batteries) must raise engine oil pressure to 6psi before the engine can be cranked. This cuts out as soon as the engine is running and oil supply is provided by the main engine-driven pump. Filtration for the lubricating oil is located above the engine at the free end and a cock is provided allowing a sample to be drawn from the lubricating oil system for analysis. A lubricating oil filler and dipstick are located on the crankcase inspection covers on the 'A' bank side.

A thermostatically controlled bypass valve is located on the 'B' side which prevents oil from passing through the oil cooler until it reaches the optimum temperature. Once the oil reaches 65°C, the oil thermostat begins to open and oil passes through the oil cooler. The thermostat is fully open when the oil temperature reaches 74°C. The maximum oil temperature is 88°C, if this is exceeded the engine will revert to idle once the secondary coolant circuit overheats.

Two cylinder blocks (one for each side of the engine) and a single crankcase are welded together to form the complete engine block.

BELOW No 43161 leads a London-bound service through the lush Devon countryside during early July 2006. Extra IC125s had been added to the First Great Western fleet as passenger demand increased. *(Steve Vaughan)*

The cylinder blocks have separate liners manufactured from chrome-plated drawn steel tube enabling these to be changed when the engine is overhauled without the need to re-machine the cylinder block itself. The main bearing caps for the crankshaft are held in place by the crankcase assembly and are cross-bolted to increase rigidity. There is also a longitudinal bar running across all the caps to restrain movement due to firing loads on the crankshaft bearings. Extra material is built into the block at the driven end of the engine to support the flywheel and alternator, which is supported by the engine at the drive end.

The crankshaft is forged from alloy steel and nitride hardened to minimise wear; the main journals and crankpins are polished to give a low-friction surface. An additional bearing and split gear are provided on the drive-end to support the flywheel and alternator, and to drive the auxiliary gear train for the lubricating oil pump.

The connecting rods attach the pistons to the crankshaft at the 'big end' using either a blade ('A' bank side) or a fork ('B' bank side). The other end is connected to the gudgeon pin at the 'small end' which sits across the inside of the piston head assembly. The fork and blade-type rods allow for a compact assembly, the connecting rods simply attach to the rod from the opposing cylinder bank. The piston body is cast from a low-expansion aluminium alloy and a cast iron insert is used at the top of the piston to minimise groove wear. The piston crown is cooled by an oil jet from the small end of the connecting rod. Three compression rings press against the side of the cylinder liner to maintain compression within the cylinder and

ABOVE LEFT The gigantic flywheel from a Valenta does not quite fit on a standard pallet.

ABOVE The engine block from Valenta engine No S231 on its side.

BELOW An exploded drawing of the crankcase and cylinder block. There are too many individual parts to detail, but the bulkiest items are two-cylinder blocks and a single crankcase.

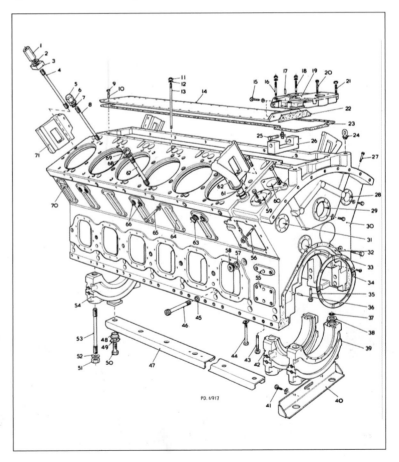

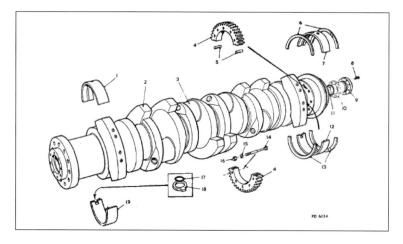

LEFT **LEFT** An exploded drawing of the crankshaft. This is designed to be a low-stress assembly to improve engine reliability.

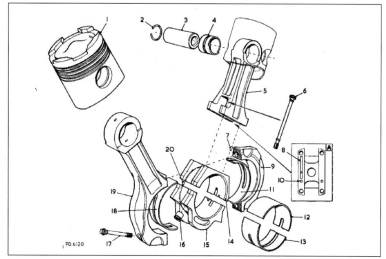

an oil control ring at the bottom of the piston skirt stops oil from entering the combustion chamber. If this becomes worn the engine will burn oil. Put simply, the piston is connected to the crankshaft by the connecting rod and this transfers the reciprocating motion of the piston into rotary motion which turns the alternator.

At the top of the cylinder is an inlet valve and an exhaust valve which allows air in and exhaust out from the combustion chamber formed between the cylinder head, piston and liner. An injector in the middle squirts fuel into the cylinder to provide combustion to the process. The cast iron cylinder head contains the combustion chamber and each head has two inlet and two exhaust valves. Each of these valves is actuated by pushrods driven by

BELOW Bearing shells both new and used from a Valenta.

BOTTOM The big-end of a connecting rod in situ on a crankshaft.

ABOVE An exploded drawing of the connecting rod assembly.

RIGHT Laid out on a table are a single piston from a Paxman Valenta engine, a blade rod and a bearing block.

RIGHT A set of connecting rods from a Valenta. This is a display at an open day; workshop conditions are never this tidy or clean!

the central camshaft. There are bridge pieces between valves as there are two pushrods for the four valves. The interconnection between cylinder head and crankcase is critical as numerous coolant and oil passageways run between the two while retaining firing pressures of up to 2,000psi. The gasket between the cylinder head and crankcase is a weak point on the Paxman Valenta and leakage results in the loss of oil, water or compression, which are all critical to engine operation.

The camshaft is a single-piece steel bar with deformed ellipse-shaped lobes machined into its length which drive the pushrods up and down, operating the valve gear. The camshaft

bearings are individual castings secured to the centre V of the crankcase by set screws. Small cylindrical-shaped tappets rest on the camshafts providing the interface between the camshaft and the pushrod which are located in cast aluminium housings. Drive gears are located at the drive end of the camshaft and the drive end tappet housing is a bulkier cast iron casting to carry the drive gear loads. A further gear is located on top of the drive-end tappet housing to drive the water pump shaft.

Four solenoids mounted on top of the governor allow the five set engine speeds to be selected. In idle or power notch 1, none of the solenoids is energised. In notch 2 governor

FAR LEFT A close-up view of Valenta rockers and injectors.

LEFT Some Valenta injectors, injector pipework and cylinder head covers undergoing attention.

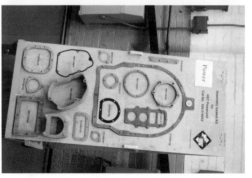

LEFT Most of the various gaskets used in a Valenta. Two have been used from this set so far.

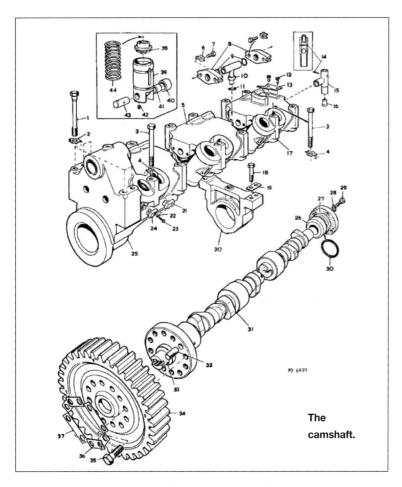

The camshaft.

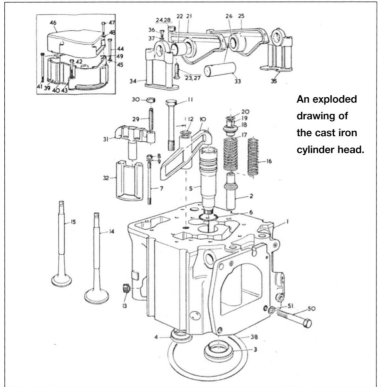

An exploded drawing of the cast iron cylinder head.

FIVE STEPS TO FULL POWER: THE POWER CONTROL SYSTEM

Unlike older locomotives, power control is not continuously variable so distinct 'notch' positions are referred to. When the power controller is moved from 'idle' to notch 1 the engine speed is maintained at the idling speed of 750rpm by the engine governor, traction motor contactors close and power output of 370hp is reached. When notch 2 is selected engine speed increases to 1,000rpm increasing power to 800hp. When notch 3 is selected engine speed increases to 1,145rpm raising power to 1,200hp. Selection of notch 4 increases engine speed to 1,350rpm raising power output to 1,790hp and finally, when the power controller is fully opened into notch 5, engine speed increases to its 1,500rpm maximum and total power is increased to 2,250hp.

Once full engine speed is reached around 2,050hp is available for traction power on a power car providing maximum output, less so for the power car providing ETS. Four train wires within the 36-way cable running through the train convey signals to the rear power car. In notch 1 none of the train wires are energised, in notch 2 train wire 10 is energised, in notch 3 train wires 10+11 are energised, in notch 4 train wires 10, 11+13 are energised and in notch 5 train wires 10, 11, 13 and 23 are energised. The engine speed responds directly to those demand wires and the load regulator demand module in the remote power car interprets the signals sent via the train wires so the power cars work in conjunction with each other.

With two power cars a maximum of 4,500hp is available to the train, easily enough to accelerate a typical eight-carriage, 415t train to 125mph. Generally, on an unrestricted route an IC125 will attain top speed in around 8-9 minutes, or after 12-13 miles.

RIGHT This is a Regulateurs Europa 1122/2G engine governor. This unit is a centrifugal flyweight-type electro-hydraulic mechanical governor which operates through a hydraulic servo mechanism with an oil pump and reservoir being integral to the governor casing. The governor maintains the set engine speed of the engine regardless of engine loading in response to the movement of the driver's power controller. It is located above the engine at the forward end, and is driven by the auxiliary drive train.

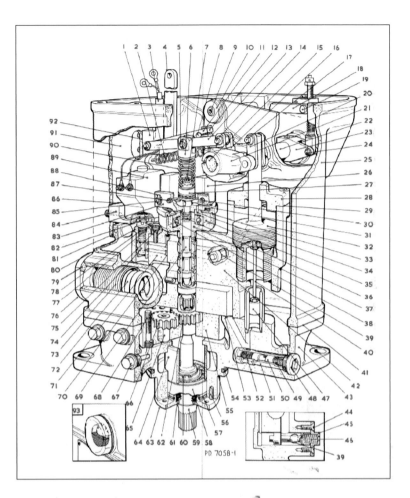

relay 4 is energised, in notch 3 relays 4 and 1 are energised, in notch 4 relays 4, 1 and 2 are energised and in notch 5 all four relays are energised. A dashpot is located between the solenoid-operated oil piston lever systems and the governor system itself, to smooth the rate of speed changes and to control its speed. In the IC125 application, this is supposed to be set such that the engine takes 23sec to go from 750rpm to 1,500rpm at operating temperature. When the silicon fluid in the dashpot is lost this can result in erratic governor behaviour.

The load control system maintains engine power at a predetermined level relative to the engine speed by controlling alternator excitation, as detailed earlier in this chapter. A Linear Voltage Differential Transducer (LVDT) is connected to the governor output position and speed setting position by an adjustable linkage. The LVDT is a transformer formed by a moveable core within three windings and is excited by a signal from the Load Regulator Demand Module (located in the electrical cubicle). The core position then indicates the loading of the engine. If the rectified output signal is showing negative then the engine is under-load. If the output signal is showing positive then the engine is over-loaded. The engine is perfectly loaded when the output is zero. The governor will always try to maintain the engine speed regardless of the load, as excitation on the alternator changes in reaction and increase or decrease in load. The governor will adjust fuel demand to maintain the required engine speed.

The governor connects to both 'A' and 'B' bank side fuel racks by a linkage system connected to the output arm of the governor.

LEFT The later 3G engine governor.

ABOVE The engine is mated with the alternator set before insertion into the power car and the whole assembly is secured within the vehicle using Cushyfoot-type mountings, a set of which are shown here. These include a rubber element to reduce transmission of vibrations from the engine to the power car underframe.

exceeds 1,725rpm. There is an overspeed reset button on top of the switch allowing drivers and engineers to see if there has been an overspeed event. Normal practice is to permit one reset, but if a subsequent trip occurs the engine should be investigated.

Also located on the 'A' side of the engine is the fuel limiter which is used on the original 2G governor (the later 3G has this built inside the governor). The fuel limiter will hold back the amount of fuel supplied to the engine as a function of the boost pressure to try to limit black smoke emissions. As the boost pressure increases so the limiter reduces its influence on the fuel rack and hence full power.

Finally, a roof-mounted electrically driven scavenging fan in the engine room removes waste heat from the engine compartment itself to keep the compartment temperature down.

Radiator compartment

The cooler group consists of banks of aluminium radiator panels through which ambient air is forced by a large fan. It is designed to keep the engine within its operating temperature limits, which are 81°C to 89°C for the primary circuit (that coolant circulates around the engine block and turbocharger) and 35°C to 41°C for the secondary circuit (that coolant circulates around the oil cooler and charge air cooler). A Vulkan coupling connects the cardan shaft which drives the cooling fan to the engine and this dampens vibration preventing damage to the engine.

There is a transverse bulkhead at the luggage compartment end of the radiator compartment which has a door on the 'A' side of the power car with a passageway through to the engine compartment. Between the engine room and the cooler group there is also a door on the 'B' side to allow access to both sides of the cooler group. The coolant hose connections are all located on the face of the cooler group adjacent to the engine, and a coolant header tank sight glass is provided which is visible from the engine compartment. Coolant capacity is 682 litres (151gal) and should be filled with BR special antifreeze to approximately a 25-30 per cent ratio with water. Coolant circulation is by a pair of centrifugal pumps mounted at the free

BELOW An unusual view of the space the cooler group occupies, showing the pipe connections in the floor for filling and draining.

The governor also incorporates the fuel shut off valve which is used to stop the engine. Spring return levers mounted on the control shafts opposite each fuel pump ensure an engine can still be shut down in the event of a fuel rack sticking open. An overspeed mechanism is incorporated below the governor on the 'A' bank side. The overspeed trip is a simple centrifugal device which trips if the engine

end of the engine. The primary pump runs at around 2.65bar and the secondary pump runs at around 1.45bar. Externally on the power car bodyshell are full-height grilles either side of the cooler group allowing air to enter the radiators. The roof of the power car is missing in the cooler group compartment as the cooler group itself provides the roof structure. Hot air is expelled through the roof of the power car. Once again, dual sourcing meant two different types of cooler group were fitted to the power cars from new. The Marston Excelsior cooler group was used in Nos 43002–43152 and the Serck Behr cooler group was fitted to Nos 43153–43198. Although of different designs the cooler groups have identical connections and functionality and are interchangeable. This soon resulted in either type of cooler group being used in any power car. There was, however, an attempt to keep all the Serck cooler groups in Western Region-based power cars for ease of maintenance, and in the privatisation era from 1997, the Serck groups were generally, but not exclusively, to be found in Great Western power cars.

The Marston Excelsior cooler group with Voith hydraulic fan drive (type 1120H/10-12) consists of four radiator panels mounted in a steel frame; two lower panels mounted vertically at the bottom of the cooler group for the secondary cooling circuit, and two upper panels mounted at 45° from the vertical at the top of the cooler group for the primary cooling circuit. Air is drawn through the power car bodyside louvres, through the secondary panels and out through the primary panels and expelled through the roof. A 90° gearbox converts the horizontal drive of the engine to vertical drive for the fan which is mounted just above the gearbox approximately half-way up the cooler group, below the primary panels. The bevel gearbox also drives a secondary cooling circuit water pump and a dual heat exchanger to cool oil from both bevel gearbox and cooling fan circuits. The fan hub consists of a fluid coupling whose impeller is connected to the mechanical input drive. The output speed of the fan is controlled up to a maximum of shaft speed by the amount of oil in the coupling; the fan speed increasing when more oil enters the coupling.

The two circuit temperatures of the coolant are measured on the engine by two thermostats

which activate the fluid flywheel via a Pneumatic Regulator Valve. A fan speed control air isolating cock is located in the engine compartment on the 'A' bank side wall. If no cooling is required the fan is held static by a compressed air brake. A header tank is provided at the top of the cooler group for both circuits and there is a single fill point for both tanks and a combined overflow. The tanks ensure that the cooler group is kept topped-up with coolant and if the level falls to such a point where the header tank is empty a float switch activates the Low Water Switch detailed below. Coolant is carried between the engine and cooler group via reinforced rubber hoses with flanged connections at each end, each circuit having separate feed and return

ABOVE The Marston Excelsior type cooler group, as fitted to the first 151 production power cars.

BELOW The top half of a Marston cooler group showing the primary panels and connections.

RIGHT Fans from four Marston cooler groups undergoing overhaul.

FAR RIGHT An empty shell from the bottom half of a Marston cooler group showing the fan mounting normally obscured by radiator panels.

RIGHT A diagram showing the panel layout and airflow on a Marston cooler group.

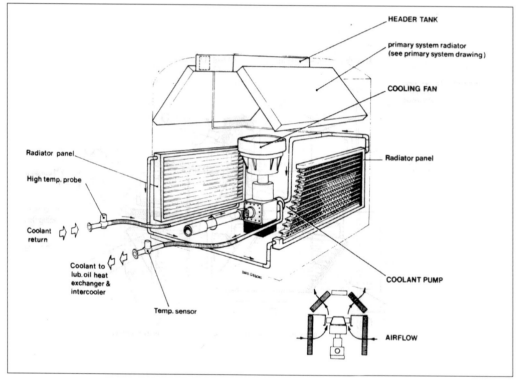

HEADER TANK

primary system radiator
(see primary system drawing)

COOLING FAN

Radiator panel

Radiator panel

High temp. probe

Coolant return

Coolant to lub. oil heat exchanger & intercooler

Temp. sensor

COOLANT PUMP

AIRFLOW

RIGHT A Marston cooler group viewed from above the power car, showing the roof slat arrangement.

hoses while smaller hoses link the cooler group with the header tank.

On the roof of the cooler group lightweight aluminium opening slats were fitted. These are forced open by the air flow and drop closed when the fan is not running. A thermostatic bypass valve prevents undercooling of the primary circuit, restricting the flow of coolant until the engine has reached normal operating temperature.

A High Water Temperature (HWT) thermostatic switch operates when the coolant temperature reaches 95°C in the primary circuit or 71°C in the secondary circuit. When this switch operates traction power is removed from the affected engine and the engine returns to idle and cannot be reinstated until the coolant

temperature has fallen sufficiently. A Low Water Switch (LWS) operates when the coolant level in the system has reduced to such a point that the engine cannot continue to operate without damage occurring. Upon operation of the switch the engine affected is immediately shut down, the low coolant light will also illuminate on the cubicle bulkhead. The engine cannot be restarted until the coolant level has been raised by replenishment.

The Serck Behr cooler group is significantly different in two main areas. The fan is driven via a hydrostatic pump rather than via a viscous fluid coupling. The fan and motor are mounted close to the roof of the cooler group. The Serck cooler uses just two larger vertically mounted panel sets on either side of the cooler group, the outer panels are for the secondary circuit and the inner panels are for the primary circuit. The air flow is drawn through the secondary panels, across a narrow gap and through the primary panels before being expelled through the roof. The primary and secondary cooling circuits themselves are laid out in exactly the same manner as the Marston cooler. A pump is mounted at the bottom of the cooler group and connected to the free end of the engine via a cardan shaft. Connections are then made from the pump to the fan motor to carry the hydraulic oil around the system with a thermostat controlling the pump output such that at low temperatures the fan motor stops completely. Fluid flow is then increased up to the maximum as the engine temperature increases. The HWT and LWS systems operate exactly the same as a Marston but the roof ventilation system is much simpler. A fixed grille arrangement is used instead of the opening and closing slats. The Serck cooler is

ABOVE This is the less- common Serck Behr-type cooler group as fitted to the last 46 production power cars, standing beside Marston-fitted No 43164 at Laira depot. This allows comparison between the roof arrangement on the two types. *(Mark Parsons)*

technically far better at doing its job than the original Marston design, which suffered from insufficient thermal performance and also had a major flaw: there was no protection system against the loss of hydraulic oil. As the oil seals became worn, oil was lost (mostly vented out over the roof), but when the group had insufficient oil the fan speed fell or stopped completely. However, this would not become apparent until the HWT switch operated which would often be too late to avoid damage. The Serck-type groups were phased out in the early 2000s when better cooler groups were developed. The Marston-type groups were also all replaced by 2010 and their replacements are detailed in Chapter Six.

LEFT A Serck cooler group viewed from above with non-standard roof mesh, which shows the fan position more clearly.

Luggage van and guard's compartment

The rear section of the power car was intended to provide accommodation for the guard and a storage area for bulky items, parcels and the conveyance of Royal Mail, which was then common on passenger trains. Power cars Nos 43002–43152 were built with guard's accommodation which was separated from the 1.5t capacity luggage area by a bulkhead with a sliding door. This led into the main luggage area where ladders for emergency evacuation of passenger between stations were strapped to the luggage area side of the bulkhead.

The guard's accommodation consisted of a tip-up seat attached to the bulkhead, lighting, a small electric stove, a heater and public address (PA) equipment. On these power cars some of the parking brake equipment was installed in a small cupboard on the 'A' side of the power car. This included the two governors, a reducing valve, an isolating cock and application/release control for the parking brake. An indicator gauge was visible through an aperture when the door was closed. This equipment was relocated to the clean air compartment on power cars Nos 43153–43198, but has remained in its original location on the earlier vehicles.

Externally, the power cars with guard's accommodation could be identified as they included windows at the rear and side of the power car in the guard's van area. It quickly became evident this area was not satisfactory for the guard as it was too noisy and far less comfortable than the adjacent carriages. Following pressure from the trade union it was decided new carriages would be introduced incorporating guard's accommodation and one of these would be included in each train. Consequently, power cars Nos 43153–43198 were constructed without any guard's accommodation and the bulkhead was omitted, increasing luggage capacity to 2.5t. Generally, the early power cars were modified with the guard's equipment removed upon overhaul during the 1980s, but the additional external windows were retained until removal much later, when additional equipment was installed and sources of bodywork corrosion were being tackled.

An emergency equipment cupboard is located on the 'B' side wall of all power cars. This locked cupboard contains emergency drawgear should the train require assistance. As the production IC125 power cars are not fitted with buffers or conventional drawgear, two sturdy bars are housed in this cupboard. The shorter bar allows an IC125 to be attached

RIGHT A sliding luggage van door shown in the open position.

FAR RIGHT The original bulkhead separating the guard's compartment from the luggage area, viewed from the guard's side.

RIGHT A general view of a luggage van in later years with all the guard's accommodation removed, windows plated over, and the lockable cage all long gone.

to a locomotive and the longer one allows a power car to be attached to another power car. Each bar attaches to a coupling eye located behind the cover under the nosecone of the power car and is secured in place using a detent pin provided.

An emergency lamp bracket is also housed in this cupboard, which slots into the nosecone cover should a temporary headlamp/tail lamp be required following a failure. Traction motor blowers Nos 3 and 4 are located at the bottom of the emergency equipment cupboard. Contactors and relays for the cab air-conditioning equipment were originally located above the emergency equipment cupboard as this was added too late in the design process to be incorporated in the electrical cubicle.

BELOW Original lighting and public address system for the guard's compartment.

BELOW RIGHT Parking brake controls as fitted in power cars up to No 43152.

ABOVE The drawbar cupboard opened, showing the short locomotive-to-power car bar on the left and the longer, power car-to-power car bar on the right.

ABOVE The short locomotive to power car drawbar is seen in use. The power car is coupled to a rigid-hook Class 08 shunting locomotive so the second, round-headed pin above the notice in the middle of the bar is removed as instructed.

BELOW LEFT Locomotives with later swing-head buckeye-type couplings cannot attach to the power car using the original bar, so an adapter piece is needed for coupling to Class 67 locomotives. Note also, the gap between power car and locomotive is extended necessitating extension pieces on both air connection hoses.

BELOW RIGHT An alternative view of the rear of a luggage van. This is a First Great Western power car with cycle stowage and the control 'brain' for the ATP equipment in the corner. The location of this box varies, depending on the original configuration and time of installation.

The units have 'system healthy' lights which can be viewed through a small window and a local circuit breaker, along with a separate 110V supply added after installation to improve reliability.

A lightweight folding aluminium ladder to access the windscreen was provided in the luggage van, stored on a bracket on the cooler group bulkhead wall. These were generally relocated to the wall beside the emergency equipment door to make way for other equipment such as electronic engine control panels. Although additional ladders for emergency passenger evacuation are provided in the guard's accommodation in the train, most operators have stowed another set behind the relocated windscreen access ladder.

All power cars were built with large sliding external van doors manufactured from GRP and above each door there is a light switch for the compartment. A driver/guard buzzer was also fitted allowing the guard to communicate with the driver using codes. A separate lockable cage was provided for the luggage area but these were progressively removed until the last few remaining examples were removed to make way for the new fire protection system.

The now largely vacant space at the rear of the power car has proved to be extremely useful as it became home to an array of additional equipment, which has been added during the service life of more than 40 years. Without the availability of this space it would have been very difficult, if not impossible, to accommodate many of the new systems required to keep pace with change in legislation and safety equipment on the modern railway. Various electronic engine monitoring and management systems have been installed to work with the newer engine types in this area, and Chapter Six looks at these. A few trial systems were also housed in this area prior to this use, but none was widespread or long-term.

Eight power cars (Nos 43013/014/065/0 67/068/080/084/123) were modified to act as Driving Van Trailer (DVT) vehicles for use in conjunction with electric locomotives. On these vehicles an additional electrical cubicle was installed centrally in the luggage van to create and decipher coded time-division multiplex signals sent down the train to a remote locomotive. The system was used from 1987 to 1989 and included an additional bank of switches and lamps on the power car desk. As part of trials No 43014 and No 43123 also had

motor-alternator sets fed from standard 850V electric train supply installed in the luggage vans. All of this equipment was later removed.

The fire system was replaced on all power cars during 2000–2002 as the previous

BELOW The red boxes in the cage are the brake air drying equipment which was fitted to 31 power cars then allocated to the Midland Main Line. This equipment was intended to speed up the rate at which the air tanks could be recharged without the build-up of water, reducing corrosion of the air equipment. The system failed to meet expectations and was soon isolated permanently.

Halon-based system could no longer be refilled following a change in legislation. The replacement system known as INERGEN (INERt Gas ExtiNguishing system) was installed on the 'A' side of the luggage van and consists of three large gas bottles and a smaller trigger bottle. Piping connects a manifold on top of these bottles to the engine room and clean air compartment – an area not covered by the earlier system. Sensors for this system are located in the engine room and clean air compartment which fire a solenoid on top of the trigger gas bottle, which then opens valves on the three main bottles.

## Exterior (front)

The ETS socket inside the front hatch allows the 415V three-phase ETS supply to be connected to a shore supply via a jumper cable, such as those found at depots and some major terminus stations. The socket is of the standard type but has two additional pins for operating a safety loop circuit which prevents traction power from being applied with the shore supply plugged in. The engines can be started with the ETS connected to a shore supply but it must be unplugged before either power car will feed ETS to the train. The main reservoir (yellow) and brake pipe (red) are standard pipes; the main reservoir pipe head has a star valve inside the head which closes the pipe when not connected to another.

The coupling eye on the power car allows two different types of draw bar to be fitted. The short draw bar has a head on the opposite end which goes on to the draw hook of a locomotive and is secured in place with a long-handled detent pin. The long bar is only for joining two power cars together and although longer it is significantly lighter than the short bar as it contains fewer parts and is thinner. The coupling eye is secured in the power car with a pin and in order to attach another train the pin should be pulled up and the bar slides out by around 9in/15cm. The assisting train is brought into position and the other end of the bar is attached to that train. The assisting train is then gently brought forward until the pin drops on the IC125 coupling eye. When all pins are in place, with detent pins locking them in position, the air pipe connections are made.

Two horns are mounted in the centre behind the grille cover, which is a largely decorative plastic cover designed to be easily and cheaply replaced in the event of damage. The horns are nominally tuned to E*b* and G*b* and are loud enough to be heard for some distance

ABOVE The imposing front-end viewed head-on from track level. Located just above the point of the nose cone is the unit containing the frontal lighting and horns. Below the horn grille is the openable hatch which is secured in place with four budget locks. The lower skirt panels around the hatch and extending below the cab sides are bolted on replaceable items, which can be quickly changed if damage occurs. *(John Zabernik)*

LEFT Eight power cars were fitted with front buffers and conventional drawgear to operate with electric locomotives for a period between 1987 and 1988. This equipment was retained and substantially changes the front-end appearance.

RIGHT Revised light clusters featuring more reliable LEDs and sealed units were fitted progressively from 2001. Maintenance of the original units was problematic as screwing fittings and covers to the glass fibre structure became troublesome over time.

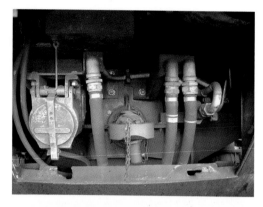

RIGHT Inside the front hatch – from left to right: train-to-shore Electric Train Supply (ETS) supply connection, main reservoir pipe, coupling eye with pin, brake pipe, main reservoir pipe, brake pipe isolation cock. A light switch and light (out of view) are also provided.

RIGHT The front with valance removed. The ETS socket has been latched open showing the plug connection.

BELOW The cab air-conditioning module in its original position underneath the nose cone, hidden behind the front fairing. It was located here to give it proximity to the cab and for ease of maintenance. On the eight buffer-fitted power cars an additional guard had to be added to protect the module.

such that a warning of a train travelling at high speed is heard early enough for track workers to move to a position of safety. They are driven using compressed air from the main reservoir pipe via the horn control solenoids. Either side of the horns are the light clusters which mirror each other. The original configuration was (from the outer edge of the power car): headlight, red tail light and white marker light. The headlight is of the automotive type rated at 100W and is driven via a transformer which is fed from the ETS supply. The tail and marker lights are of standard BR design using 110V low-wattage lamps changed by hinging the coloured cover forward. The lights are located behind plastic covers which are secured in position with aluminium surrounds, although often following replacement of the plastic covers, the surround was omitted. The lights have since been changed for lower maintenance sealed units using LED lights for tail and marker lights and improved high-intensity headlights. Midland Mainline power cars initially received new units which (from the outer edge of the power car) were changed to white marker light, headlight and red tail light, but when all the other IC125 operators followed suit they used a simplified arrangement whereby the tail and marker lights were incorporated in the same fitting.

The cab air conditioning module contains a compressor, liquid receiver and condenser fan and coil, together with associated control gear, solenoids and pipework. The compressor causes the gas to turn to liquid which is passed though hoses connected to the unit to the evaporator mounted behind the desk in the cab. Here, the liquid turns to gas as it removes the heat from the air and passes back to the air-con module where the heat gained is removed via the condenser coil with the aid of a fan. The hoses connecting the unit to the cab are removable as are all the electrical connections which run from the 415V three-phase ETS supply. More recently, a smaller, more reliable module with an adjustable thermostat has been developed and this has been relocated to inside the cab itself in front of the second seat. This significantly reduces the leg room for the occupant of the now-infrequently used second seat, but protects the module from damage.

Exterior (bogies)

The power cars sit on two BP10 bogies which are a 'Bo-Bo' arrangement with two driven axles per bogie. The bogie design was born out of the developments striving to keep the un-sprung mass to a minimum. The bogie has a welded steel frame with 1,020mm diameter Monobloc wheels with cast iron cheek plates upon which the detachable brake disc linings are mounted on both inner and outer faces. There are two classifications of bogie:

the BP10B has Brush traction motors fitted while the BP10C has GEC traction motors fitted. Other than the motors and associated gearboxes the two types of bogie are identical.

The bogie is mounted to the body through the centre pivot bearing with primary and secondary suspension mounting directly to the frame at various points. The two bogies are identical except that the front one is fitted with lifeguards ahead of the leading wheelset to deflect debris. It also has the AWS/TPWS receiver mounted at the rear to relay signalling

ABOVE No 43384 is seen speeding through Natton, near Ashchurch, Gloucestershire on 27 March 2009 while operating 1V19 06.00 Leeds to Plymouth, with No 43207 at the rear. *(Jack Boskett)*

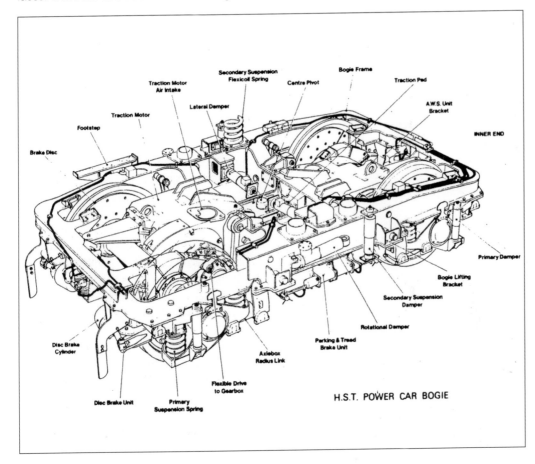

H.S.T. POWER CAR BOGIE

LEFT This wonderfully detailed drawing shows the BP10 bogie in all its glory.

RIGHT A BP10 bogie underneath a power car. This one has been painted for an exhibition, which usefully picks out some features such as the dampers including the horizontal yaw damper between the bogie and power car body. The pair of three-handled parking brake release handwheels, and the primary suspension springs are all painted white. Axle box covers are in yellow while to the right of the picture, filler points for coolant can be seen picked out with blue paint.

indications to the cab. Primary suspension consists of large diameter coil springs supporting the axlebox castings within which are the retained axle bearings. This primary suspension accommodates the large excursions seen by the wheelsets. The yaw dampers can

be seen in the middle of the bogie around platform height – all of the hydraulic dampers absorb energy and limit movements.

There are also four secondary suspension coils, two per side of the bogie, which support the underframe from the bogie frame and these accommodate the lesser movements which occur between the body and bogie.

The power cars provide a large proportion of the brake force (35 tonnes each) for the complete IC125 train set, using both disc

BELOW The BP10 bogie viewed from the other side. This is the rear bogie on a Great Western power car with the Hasler speed probe for the ATP equipment visible on axle three.

RIGHT Primary braking on the power car is by disc brakes. The brake actuator moves outwards to press the brake pad against the disc. The exposed position of the brake actuator led to issues during snowy conditions, a build-up of ice and snow around the actuator can critically reducing its ability to operate. In such conditions, drivers are instructed to make regular brake applications to prevent ice build-up.

brakes and tread brakes. The disc brakes are similar to those used on the trailer cars with one pneumatic actuator per wheel closing a calliper down on two brake pads. The discs on the wheels are changeable should they become too thin, without the need to change the wheelset. There are also tread brakes, one per wheel, which operate via air pressure in the combined Parking and Tread Brake Unit (PTBU), which is mounted under the rotational damper on each side of the power car. This dual method of braking provides the power cars with sufficient stopping power (via the discs) as well as a method of scrubbing the wheels, to remove any railhead contamination, aiding adhesion. The tread brakes also act as parking brakes and are actuated via the same PTBU. The spring-loaded unit applies the cast iron tread brake blocks against the tyre and they

LEFT The tread brake which acts as the parking brake and a scrubber to clean the tyre during braking.

RIGHT A bracket on the leading bogie houses equipment associated with the signalling system. The AWS/TPWS receiver (see cab equipment) is located just behind the bogie frame. The yellow receiver hidden ahead of this indicates this is a Great Western power car with ATP equipment. The receiver picks up signals from track-mounted equipment as the train passes over it.

FAR LEFT The standard axle end bearing on a BP10 bogie.

LEFT An axle end bearing for the wheelslip protection system on a BP10 bogie.

RIGHT A Brush TMH 68-46 traction motor mounted in the bogie frame. Four of these are fitted under most IC125 power cars. The yellow piece is a temporary cover where the traction motor blower would normally be connected. The cables feeding the traction motor can be seen towards the top of the photo.

can only be released by the application of air pressure. Manual release handwheels for the parking brakes are fitted adjacent to each wheel allowing the parking brakes to be wound off in the event of a defect in the air system which prevents them being released normally.

The traction motor was another item which was dual-sourced when the IC125 was constructed. Power cars Nos 43002–43123 and Nos 43153–43198 were fitted with Brush TMH 68-46 motors while Nos 43124–43151 were fitted with GEC G417AZ motors. Both machines are of a DC,

RIGHT A Brush traction motor showing the driven end.

FAR RIGHT The rear of a Brush traction motor.

RIGHT A gearbox cardan shaft

FAR RIGHT Gearbox flexible links in situ.

series-wound four-pole design (multiple brush sets) and are of similar construction.

DC motors consist of an armature which rotates on a pair of bearings (one at each end) suspended within a cast magnet frame containing a set of field coils. The armature is constructed from a magnetic steel core stamping pack with imbedded coils formed of insulated copper bar. Current reaches the rotating armature using shaped carbon brushes. These are spring loaded against the copper commutator on the shaft assembly. The armature coils are connected to the commutator segments and switch the appropriate coils into circuit as the shaft rotates. The brushes wear out after a period and require changing at intervals. The motors are wired in series-parallel, each bogie has its traction motors wired in series, then the two bogies are wired in parallel. The traction motors also have speed probes fitted to detect the speed of rotation and these provide a signal for the speedometer and for the Wheel Slide Protection (WSP) unit detailed earlier in this chapter. More recently, these have been superseded by axle-end probes on the later Brush electronics which employs a Westinghouse WSP system. The new probes allow for additional monitoring function such as a 'locked axle' indication as well as feeding data to the OTMR system.

Extra sensing cables also allow for detection of wheelslip by comparison of motor voltages via a wheelslip detection circuit and the WSR relay is triggered if the voltage across one motor increases relative to the other, indicating a slip. The motors are ventilated via two AC traction motor blowers mounted in the power car body. One of the unique features of both traction motor types is that they do not require weak fielding at higher speeds. This results in a far simpler and more reliable control system. The traction motors drive the wheelset through single reduction gears via a flexible coupling. The couplings used consist of flexible links that allow motor/gearbox an amount of movement relative to the wheelset. The motor/gearbox assembly are supported on the frame of the bogie via radius links mounting on to the front of the bearing caps as well as rubber block mountings on the top side of the bogie frame.

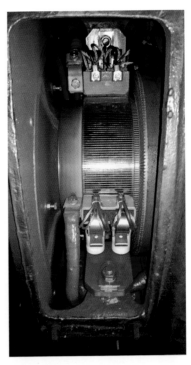

ABOVE A torque reaction link.

ABOVE A traction motor commutator.

BELOW A traction motor blower. These are housed in the body of the power car below the rectifier bank and below the emergency equipment cupboard in the luggage van.

British Rail had developed an almost pathological fear of being over-charged for components and this resulted in a policy of dual sourcing as many components as possible. The idea was that by using two different suppliers they would get a better price when buying equipment and consumables, but this policy was to become something of a nightmare for the IC125 engineer and created reliability issues which remained throughout the whole life of the fleet.

Once an item had been successfully designed, purchased, installed and tested the whole process then had to be repeated for the same item from an alternative supplier! The fleet was then saddled with the problems this created which ranged from staff familiarity with the different components to the cost of holding spares for each different item. There were four major items on the IC125 power car which were dual-sourced: cooler groups, brake pipe pressure control units, compressors, and traction motors. To prove the secondary sourced item would function correctly some power cars received different equipment for trials during early operation. The first power cars to receive the Serck-type cooler groups were Nos 43026 and 43027 during an overhaul in March 1978. Apart from a few months while one cooler group at a time was modified, they retained the Serck cooler groups until December 1981. In common with most of the other dual-sourced items the cooler groups were just as reliable and entirely interchangeable, so after the first overhaul cycle they appeared in power cars randomly.

The traction motors were a very different story and in retrospect dual sourcing of the traction

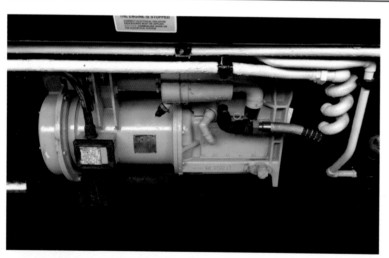

Exterior (between the bogies)

Compressed air is needed primarily for the braking system but on the IC125 it is used on other items such as carriage suspensions and other various ancillaries. Once again, the air compressor is a dual-sourced item with power cars Nos 43002–43152 fitted with D&M compressors and Nos 43153–43198 with the alternative Westinghouse units. Both are three-stage reciprocating compressors and are completely

ABOVE LEFT Davies & Metcalfe (D&M) air compressor as fitted to the majority of power cars when built. This is mounted on the 'A' side of the power car, behind the front bogie. The compressor feed runs via the brake frame and charges the air reservoir tanks located underneath the rear of the power car.

LEFT A D&M compressor viewed from the 'B' side, also showing the engine bedplate waste drain cock and the fuel pump.

OPPOSITE A D&M compressor on a pallet awaiting overhaul.

motors for the IC125 turned out to be a huge mistake. No 43018 received GEC motors in April 1979 for one year, but it was not until the fleet for the West of England services began to enter service that serious problems with the GEC traction motors became apparent. The GEC motors simply were not robust enough; commutator bars were distorting and this was significantly shortening the life of the traction motor brushes. The new fleet could not enter service until the problems had been resolved and several months during 1980 were spent improving the manufacturing process and materials used before the GEC motors could be passed for operation. Traction motors for the North East to South West route fleet were also to have been supplied by GEC. That was hastily changed creating a further delay while Brush motors were produced instead.

No 43126 entered service with Brush motors (presumably using spare stock) allowing just that one power car to enter traffic ahead of the rest of the batch. It retained Brush motors until 1983. Nos 43151 and 43152 also ran initially with Brush traction motors until enough of the GEC motors had passed reliability tests.

Although the bogies are interchangeable between vehicles, in later years the GEC motors were generally confined to power cars Nos 43124–43152 enabling them to receive special monitoring and attention owing to the excessive commutator wear. From 2002, they were allocated to Landore (Swansea) depot for this purpose and continued to work in the same fleet and from a single depot until their reallocation to Haymarket (Edinburgh) for ScotRail services.

LEFT A Westinghouse compressor as fitted to Nos 43153–43198 from new.

interchangeable, so were soon swapped around between power cars after overhauls began. However, for ease of maintenance, the Westinghouse type was generally confined to Western Region-allocated cars.

A large, three-phase induction motor at the front drives a three-cylinder compound compressor at the rear, which feeds air through an intercooler mounted in front of the fuel tank. A third, revised-type of compressor also supplied by Westinghouse, was later fitted to some power

LEFT A modified Westinghouse compressor with a new motor, as fitted to some power cars from 2003 onwards.

BELOW On 11 September 2011, Grand Central's 18.24 London King's Cross to Sunderland is seen making an unusual stop at Grantham with No 43423 *Valenta 1972-2010* leading the train. This power car was named to mark the demise of the last of the original engine type, this being the final production power car to operate with the classic engine in normal service. *(John Zabernik)*

RIGHT The battery isolation switch and shore supply box to the left of the main (No 2) fuel tank with fuel gauge (showing only one-quarter full), fuel filler and sight glass mounted on the right-hand edge. The brown-coloured pipes/fillers indicate they carry fuel whereas the filler located to the top right of the fuel filler is coloured pink indicating this is for engine oil. Air pipes running the length of the power car are painted white. It should be stressed this power car was turned out like this for show purposes and painting such items is not normal practice.

cars and reliability of the compressors was later improved by the installation of local air filters reducing the amount of pipework to the air inlet. This was previously located on the 'B' side of the power car, towards the rear.

The IC125 power cars were originally designed to be fitted with three fuel tanks: No 1 tank mounted on the 'B' side of the power car, opposite the compressor (1,002 litres/220gal); No 2 tank mounted across the power car width behind the compressor and in front of the battery boxes (2,739 litres/602gal), and No 3 tank mounted between the battery boxes (1,941 litres/427gal). This gave a total maximum fuel capacity of 5,682 litres (1,249gal).

LEFT The No 3 fuel tank is concealed between the battery boxes.

LEFT Battery boxes with one half dropped open for maintenance and the isolation switch operated.

ABOVE Inside the shore supply box showing the connections and the switch to change from shore supply to battery supply for internal lighting.

Early experience with the production power cars showed that the inclusion of No 1 tank upset the balance of the vehicle with a tendency for it to lean towards the 'B' side when the tank was full. Consequently, only power cars up to No 43037 were built with the No 1 fuel tank. These earlier power cars generally lost their No 1 tank during overhaul, although No 43012 had the tank removed during repairs to collision damage. Since the removal of the No 1 tank, the maximum capacity was reduced to 4,680 litres (1,029gal).

Industrial grade 'red' diesel is used in common with other non-road applications. The fuel tanks are interconnected with a 76mm diameter flexible hose which allows only the No 2 tank to have the fuel gauges fitted, by virtue of the level in the tanks equalizing. The fuel filler is located at the top of the No 2 fuel tank on both sides of the power car. A Bayham gauge is fitted in the centre on both sides of No 2 tank giving an indication between empty and full and additionally, a sight glass-type gauge is also fitted to the side of No 2 tank to allow a visual check of the actual fuel level within the tank. An electrically driven Varley pump is mounted on the 'B' side of the power car just in front of No 2 fuel tank which pumps fuel up into the engine compartment. The pump has a DC motor as it must run off the battery supply before the engine starts to allow priming of the engine.

The batteries are divided into two sets of battery boxes with a pair of inter-joined boxes each side of the No 3 fuel tank. There are 12 cells in each battery box each with a nominal rating of 2.3V with four sets totaling 48 cells which results in a nominal 110V when on charge as the cells are connected in series. Battery voltage is normally 96V and must be

BELOW The 09.52 Aberdeen to London King's Cross dashes through Peterborough on 3 August 2013 with No 43318 leading and No 43312 at the rear. The train still wears the silver and white National Express livery, albeit rebranded with East Coast names. *(John Tattersall)*

in excess of 86V for the engine to be started. The batteries are then connected to the Battery Isolator Switch (BIS) which is interconnected so it can be operated from either side of the power car. This is located just forward of the battery boxes. In the horizontal position the batteries are isolated and in the vertical position the batteries are connected.

The BIS supplies all DC machines apart from the lighting circuits and fuse tester via the Shore Lighting Switch (SLS). When the SLS is in the 'battery' position lighting power is taken directly from the battery. When in the 'shore' position lighting power is taken from the shore lighting socket and the battery is fully isolated (provided the BIS is 'out'). The SLS and charger socket are located below the BIS under a small sliding flap and on the 'B' side only. This cover also contains a shed lighting socket. Located to the rear of the batteries are three pipe connections, these are repeated on both sides of the power car. The top connection is the coolant filler with a self-sealing coupling, the middle connection of which is a drain for the secondary circuit with an adjacent overflow, and the bottom connection is the primary circuit drain.

Exterior (rear)

As built, a single pair of main reservoir and brake pipes were provided between the power car and the first trailer, but between carriages these pipes were duplicated allowing for a defective pipe to be isolated in the event of a breakage or leak. Power cars were retrospectively fitted with dual main reservoir and brake pipes. Two ETS connectors on the rear of the power car house the cables that supply the auxiliaries for the train, such as

ABOVE A general view of the rear of power car No. 43239 showing its gangway connection to the train and security door. Below the gangway connection is the Alliance coupler connection. As this is an East Coast power car it has an additional communication cable located to the right of the coupler. A tail light has been put on the hook ready for this power car to make a solo trip on the mainline. *(John Zabernik)*

RIGHT The top of the rear section of the power car showing the structure underneath the glass fibre moulding while removed for corrosion repairs. This view also shows a locomotive-style data panel as applied to a power car. The RA index of 5 indicates they have a very good 'Route Availability' of less than 19t per axle. The scale runs from 1 to 10 with 10 as the heaviest and most restricted locomotives. Most modern locomotives have a RA index of 6 or 7.

RIGHT The bottom of the rear of the power car showing the ETS supply sockets on the outer edges, dual main reservoir (yellow) and brake (red) pipes, and between those the alliance buckeye coupler. Below the coupler is the 36-way control cable socket and behind that are the air tanks running the width of the power car. Hidden from view under each tank is a bleed valve which drains any condensed water which may have formed in the tank, to prevent them from rusting.

heating and lighting as well as feeding certain functions on the other power car.

The 36-way cable runs the length of the train. A cable harness runs through each vehicle and these are connected using jumper cables between vehicles. This carries the control and communication signals except CDL (Central Door Locking), which is a retro-fitted item using its own cable. The cable layout is as follows:

1 Start engines
3 Stop engines
4 Power hold
5 Power pick-up
6 Power control (notch one and above)
7 Forward
8 Reverse
10 1st engine speed increase (notch 2)
11 2nd engine speed increase (notch 3)
12 Cab desk control positive
13 3rd engine speed increase (notch 4)
15 Brake pipe pressure control unit lines
16 Fire alarm
17 General fault indication
18 Engine stopped indication
20 Brake pipe pressure control unit lines
21 Brake pipe pressure control unit lines
22 Wheelslip indication
23 4th engine speed increase (notch 5)
28 Parking brake interlock
34 Train lights ON, driver/guard and public address
35 Train lights OFF, driver/guard and public address
36 Train supply on

RIGHT The new, lightweight 36-way cable socket on the power car.

RIGHT The original 36-way jumper cable which carries signals down the train. This is the female end which slots into the male end on the vehicles.

RIGHT Alliance-type fixed-head buckeye coupling after separation from a power car. As the trains operate in fixed formations, drop-head buckeyes, conventional buffers and coupling hooks were not considered necessary and were not fitted to the power cars or trailer vehicles. An additional retro-fitted pin keeps the coupling closed following early critical in-service failures of the coupling.

LEFT A close-up view of one of the ETS sockets, the circuit cannot be switched on if no cable is present. Two duplicated sockets (one either side) carry the three-phase train supply via heavy duty male to male fly leads.

Wires 9, 14, 19, 25, 26 and 27 are all negative and are bonded together. Wire 25 is the negative for the brake controls whereas the others are non-specific negative returns. Wires 2, 24, 29-33 were spare and some items have subsequently been added to these such as wire 31 which is now used for a remote low-coolant indication on certain fleets.

Bringing an Inter-City 125 into service

As with any type of train, a full preparation is carried out (colloquially known as a 'Prep' as per the abbreviation shown on traincrew diagrams). This consists of a number of methodical steps which encompass checking all the key equipment on the train. Traditionally, this was always carried out by a driver prior to the train entering service for the first time each day, although a further 'prep' would be required if the train was left entirely shut down and unattended for a period of time, or if it was coming back into service after maintenance attention during the day between passenger duties. Although drivers are still trained to carry out this procedure, it is now more commonly undertaken by depot-based staff primarily in an effort to extract the most productive use of a driver's time.

The 'prep':

1 Approach the power car which will be leading for the first movement. Ensure there are no 'Not to be moved' boards attached. Commence prep from right-hand side luggage compartment door from the rear of the power car.

1.1 Walk along right-hand side and check no pipes/cables attached, the battery switch is closed, no obvious signs of damage or leaks, parking brake manual release spokes are in place and external fire glass is intact.

2 Enter driving cab by right-hand side door.

2.1 Working from left to right across the desk check that the parking brake is applied, turn all external lights on, test the fire alarm, ensure the emergency brake plunger is in the raised position, check brake controller is not in emergency, and check the internal fire glass is intact.

2.2 On the fixed fire equipment check the firing pin is not raised, the transporter bolt has the tapered end showing, and the electrical connection is secure. (This relates to the original fire system, with Inergen only the healthy light needs to be checked.)

2.3 Check the emergency cupboard contains ten detonators, two track circuit operating clips and two red flags.

2.4 Check the portable fire extinguisher has not been used.

2.5 Check the repair book contains no outstanding defects which would affect the running of the train.

3 Walk through the clean air compartment into the engine room.

3.1 Check that the local engine stop button is out, the engine overspeed button is not protruding and there are no obvious defects.

3.2 Leave the engine room and enter the clean air compartment. Check that no isolation cocks on the brake frames are in the isolated position and check that the AWS and DSD are sealed.

3.3 Pass around the electrical cubicle and check the high-tensile doors are secured shut, no circuit breakers are tripped, all isolation switches are in the 'normal' position, the low water level light is blue, all fuses and spares are in place, and that none of the flagged relays are tripped.

3.4 Return to the cab and alight by the right-hand side door and pass around the front checking that all marker and tail lights are illuminated and the nose cupboard is closed and secure.

3.5 Continue down the left-hand side of the power car checking items as per 1.1.

3.6 Continue to walk down the outside of the train visually checking that between vehicles the coupling appears secure, all electrical connections are correctly connected, and all air pipes are connected and the cocks open. On each coach, check the brake distributor and reservoir isolating handles are not in the 'isolated' position, the doors appear closed and secure, all side panels appear closed and secure, and there are no other obvious defects.

4 On reaching the rear power car:

4.1 Repeat the steps in 1.1 and check the electrical shore supply is disconnected at this end.

4.2 Enter the cab by the left-hand side door, and once in the cab repeat all stages 2.1-2.5 and 3.1-3.3.

4.3 Move the brake controller to 'emergency', open the desk and start both engines. Then switch train supply on at this end, check the battery charge ammeter is showing a charge and the three red phase lights are illuminated. Leave the cab and check the tail lights, headlights and marker lights are all illuminated. Return to the cab.

4.4 When the main reservoir pressure registers 6.5bar turn the parking brake test switch to 'test', move the brake controller from 'emergency' to 'running' and observe the brake pressure rise and settle at 5.1bar, then move the brake controller from 'running' to 'emergency' observing the brake pipe pressure fall to 'zero' and the bogie brakes rise to their maximum of 4.65bar. Return the brake controller step by step to the running position observing the brake pipe pressure rise to 5.1bar and the bogie brakes fully release and leave the brake controller in 'running' for approximately one minute. Press the DSD pedal and move the master switch to forward then release the DSD pedal and observe the brake pipe pressure fall towards 'zero' after 2-3sec and the bogie brake pressure start to rise. Press the DSD pedal again and observe the brake pipe pressure rise, return to 5.1bar, move the brake controller to step 2 and release the parking brake. Turn the test switch to 'normal'.

4.5 Move the master switch to 'off' to close down the desk and remove the key. Switch off marker and headlights and check the windscreen wipers and washers.

4.6 Alight from the cab ensuring both cab doors are properly closed, then check that only the tail lights and no other lights are left illuminated on the rear of the train.

4.7 Proceed down the right-hand side of the train checking items as per 3.6.

5 Re-enter the leading driving cab.

5.1 Repeat the cab checks as per 4.4.

5.2 Switch off tail lights and check the windscreen wipers and washers.

Once the 'prep' has been completed a brake continuity test needs to be carried out. In order to do this the driver in the leading cab will apply the parking brake, move the brake test switch to the 'test' position and move the brake controller to 'running'. A second person (usually the guard) will operate the emergency brake from the rear cab by striking the plunger or moving the brake controller to the 'emergency' position. Both will observe the brake pipe fall to zero bar before the person in the rear cab releases the emergency brake, then both will observe the brake pipe pressure rise to 'release' (5.1bar). This proves the brake is working continuously throughout the whole train, and is the final test before the train can operate.

RIGHT NSW-TrainLink continues to use its fleet of XPT power cars intensively. The fleet had its original Paxman Valenta engines replaced by Paxman VP185 engines shortly after the turn of the millennium, and they continue to give reliable service. The fleet is planned to be replaced as they are now considered life expired.
Power cars Nos 2003 and 2006 (nearest the camera) are seen powering away from the station stop at Broadmeadow on 31 March 2017, near the beginning of their 14-hour, 613-mile journey on service NT31, the 14.41 from Sydney to Brisbane. *(John Tattersall)*

ABOVE The success of the British Inter-City 125 design resulted in the design being exported to Australia where a fleet of 19 power cars was constructed locally under licence. They bore a close physical and technical resemblance to the original design, but with some features to make them better able to cope with local operating conditions, such as larger cooler groups. They were christened XPT (eXpress Passenger Train) and put to work on a series of long-distance routes radiating from their base in Sydney, New South Wales, reaching Melbourne, Brisbane and Dubbo. Power cars Nos 2000 and 2009 are seen working the 07.11 Sydney to Casino at Strathfield on 6 November 2013. The passenger carriages are not similar externally to the British Mk3s being of a local design, although there are many identical internal features. *(John Zabernik)*

Chapter Three

Maintaining the Inter-City 125

The Inter-City 125 is a train which has been required to work at the limits of technology and cover far more distance on a daily basis than any previous train. The skill of the depots and their staff in overcoming unforeseen problems and maintaining train reliability became key to the success of the world's fastest diesel train.

OPPOSITE Nos 43066 and 43043 meet nose-to-nose at Neville Hill during the night on 14 January 2018. While most people are asleep the depots are busy getting trains ready for the following day's service. *(John Zabernik)*

RIGHT A brand-new
train, brand-new
depot. No 43004 with
set No 253002 stands
beside the new shed
at St Philip's Marsh,
Bristol on 15 April
1976. (Terry Nicholls)

The advent of fixed-formation trains created the need for depots that could easily maintain the new, intensively used Inter-City 125s without the need to split the trains up on a routine basis. Previously, separate depots dealt with locomotives individually, while carriage depots carried out the more basic servicing needs of the coaching stock fleet. Ahead of the introduction of the IC125 fleet considerable investment in depots was required to enable the new trains to be quickly and efficiently serviced overnight, allowing the maximum number of trains to be deployed during the day.

Home depots

On the Western Region, two depot locations received new sheds long enough to accommodate the nine-vehicle trains, allowing both power cars to be refuelled and receive routine maintenance attention simultaneously with the intermediate carriages

RIGHT Trains arrived
at their home depot
randomly, formed with
the latest carriages
off the production
line. It was down to
the depots to marshal
them into their correct
formations. This is the
new Victoria Sidings
at St Philip's Marsh in
Summer 1976.
(Terry Nicholls)

receiving daily checks and servicing. A new depot was constructed at St Philip's Marsh (Bristol) and the existing depot at Old Oak Common (West London) was modified to accept the new trains. Allocation of the fleet was split between the two depots, but there was no requirement for a set to return to its 'home' depot for daily servicing.

Both depots were equipped with three-road servicing sheds. In terms of overnight servicing, each set would be shunted into the shed to

ABOVE Normally, the depots are only full of IC125s stabled for their next turn of duty during the night, but the two-day Christmas break allows them to be seen in this environment by daylight. This is Victoria Sidings at St Philip's Marsh depot on 26 December 1987 with six Western sets in a mixture of then old and new Inter-City liveries. *(Steve Hampton)*

BELOW While the other sets have gone out to work, No 43146 has been left behind on Embankment Sidings at Laira depot on 29 January 1989. *(Kevin Daniel)*

be refilled with diesel and the other fluid levels checked and topped up as required. Every two or three days, a basic A-exam is carried out which includes checking brake discs/pads and running gear. Trains are moved between the servicing sheds and the carriage sidings where the passenger saloons would be cleaned and prepared ready for the next day, generally passing through a carriage wash plant either before they enter the depot or between stages of their depot visit. Spare power cars and carriages would be swapped in and out of sets during the night in response to maintenance needs, or to allow time to attend to defects which the limited overnight time window would not allow.

The nightly movement of complete IC125 sets and individual vehicles within the bigger depots while working against the clock, was once memorably likened to a huge game of chess. This was often made more complicated by last-minute failures and problems discovered after the plan had been formulated. By contrast, the day shifts on the depots would be much calmer as depots would be almost empty with just a small number of IC125 sets remaining to receive unplanned repairs or higher level exams, which are carried out less frequently.

Ahead of IC125 introduction on the East Coast Main Line (ECML) four depots were similarly upgraded to accommodate ten-vehicle sets. These were Bounds Green (North London), Neville Hill (Leeds), Heaton

LEFT Parked spare at its home depot at Neville Hill we see No 43050. Note the recent replacement of the cab door has interrupted the livery lines somewhat. *(John Zabernik)*

(Newcastle), and Craigentinny (Edinburgh) with each depot receiving its own allocation. A smaller facility was provided at Clayhills (Aberdeen) for overnight servicing, but with a shed only long enough to accommodate one power car at a time the train required shunting for both power cars to be attended to.

When West of England to London services were turned over to IC125 operation, a new facility was installed at Long Rock (Penzance) to carry out A-exams, although for heavier work the fleet was initially dealt with by the two existing Western Region depots. The increase in Western Region duties also resulted in Canton (Cardiff) undertaking A-exams and overnight servicing to reduce the workload at St Philip's

Marsh and Old Oak Common. However, by mid-1983, Canton ceased carrying out IC125 maintenance and was reduced to a stabling location only until losing all involvement a few years later.

Landore (Swansea) and nearby Maliphant Sidings undertook overnight fuelling and

BELOW Turntables for locomotives went out of use following the end of steam traction – or at least that's what they thought! The advent of single-cab power cars saw the turntable at Neville Hill become highly useful. Spare power car No 43060 can be quickly turned depending on which end of the train it will be needed.

The other depots have to send power cars which require turning, on a short trip around their local area to effect a turn around. *(John Zabernik)*

Heaton's role as an IC125 depot was revived during the ten years when Grand Central operated its fleet based there. No 43465 is seen in the shed undergoing attention on 13 July 2017. *(James Trebinski)*

cleaning for sets finishing in South West Wales with Landore's role increased during the 1990s. Laira (Plymouth) was upgraded to accommodate the North East to South West route IC125s, and when the depot extension was completed, it also took on the role of servicing sets for the London services. The nature of the North East to South West route duties meant trains allocated to that route could also receive overnight servicing at Neville Hill or Heaton, and later at Craigentinny. When some Midland Main Line services were turned over to IC125 operation in 1982, the cost of a new depot was not justified and sets operating these services used depots just off the route, at

ONE VERY LONG DAY

The longest daily duty that any IC125 set has worked was East Coast diagram EC011, operating in 2017/18, which worked 5E11 06.25 Clayhills Depot–Aberdeen, 1E11 07.52 Aberdeen–London King's Cross, 1D21 16.06 London King's Cross–Leeds, 1A48 18.45 Leeds–London King's Cross, 1N35 22.00 London King's Cross–Newcastle (via Knottingley and Church Fenton), and 5N35 02.26 Newcastle–Heaton Depot. That is a total of 1,169 miles on one tank of fuel. After a brief stop for fuel and servicing, the set was on the road again by 06.00.

Bounds Green and Neville Hill instead.

As deployment of IC125s on the Midland Main Line route increased the depot at Etches Park (Derby) took on the role of overnight servicing but it did not receive its own allocation of passenger IC125s until 2018. When some Western Region sets were extended to eight carriages, only Laira was equipped to accommodate the longer trains. Old Oak Common was extended by 1986 to deal with eight-carriage sets and similar extension work was carried out at St Philip's Marsh when the rest of the fleet was lengthened.

Upon completion of the ECML electrification scheme the depots at Bounds Green and Heaton lost their allocation of IC125s, while Craigentinny gained an enlarged allocation to operate long-distance CrossCountry services now operated by IC125s cascaded from the ECML route. The new CrossCountry routes added Longsight (Manchester) to the list of depots carrying out overnight servicing. This depot was to receive its own small allocation from 1995 until 1998 when trains for the London Euston to Holyhead line services were based there.

Meanwhile, on the Western Region changes over the years saw Old Oak Common relinquish its allocation of IC125s for several years during which the fleet was split between St Philip's Marsh and Laira. Landore received an allocation of IC125 power cars (but not trailer sets) from 2002 enabling that depot to specialise in

looking after those power cars fitted with the still-troublesome GEC traction motors. From 2006, St Philip's Marsh was adapted to carry out servicing for the local train fleet and lost its allocation of IC125s. This concentrated all the trailer sets for the First Great Western fleet at Laira while power cars were split between Laira, Landore and Old Oak Common. Heaton regained a small allocation when open access operator Grand Central stationed its three IC125s there.

Introduction of new trains on the lines from Paddington from 2017 resulted in the cascade of vehicles to form shortened IC125 sets for use on internal trains in Scotland and the power cars were allocated to Haymarket (Edinburgh) resulting in the Scottish capital having two IC125 depots.

Major works

While the six initial depots were equipped to carry out exam levels 'A' to 'D', all heavier maintenance was to be carried out at Derby. Initially, each set would be despatched to Derby annually, but this was quickly extended to a visit once every two years. Between works visits the sets would cover approximately 450,000 miles dependent on duty cycle. Power cars were dealt with by Derby Locomotive Works while the trailer sets would return to where they were constructed at Derby Litchurch Lane for attention.

Power cars entering Derby Works would receive either a 'light', 'intermediate' or 'general' overhaul. Generally, the 'light' overhaul would be scheduled to take around two weeks and involve a power unit and bogie change with minor attention to the rest of the vehicle. During every other works visit the power car would receive the 'intermediate' overhaul taking four or five weeks and comprising a full bodywork overhaul and repaint along with exchange of all the major components.

The first 'general' overhauls became due in 1984 and these typically took five or six weeks, such overhauls being undertaken instead of every third 'intermediate' overhaul. Derby Works would remove all the components planned for replacement and install newly overhauled items in their place to reduce the amount of time each power car spent in the works. The removed components would be overhauled for installation in a subsequent works visitor. Prior to being sent back to their home depot power cars were given a short back-to-back test run, followed by a full formation trial, usually involving a trip from Derby to Darlington and back.

Not all major component changes during this era necessitated a trip to Derby Works and back. As early as 1976, a power unit was changed outside St Philip's Marsh depot using a steam crane and a lifting beam borrowed from Crewe, following a turbocharger failure in service. Engine lifting cranes were subsequently installed at selected regional depots, typically

ABOVE Neville Hill's lifts are supporting the weight of No 43123 during a bogie change on 10 March 1988. This was one of eight power cars to receive a modified front end including buffers. This enabled a locomotive to attach more easily to the train during the short spell when these were used as driving van trailer vehicles. No 43123 was one of the first two modified and carried extra 850V DC Electric Train Heating sockets. One of these is seen to the right of the buffer nearest the camera, which allowed a locomotive to provide power to a standard rake of carriages through this vehicle. (Paul Corrie)

LEFT No 43118 stands mid-way through a general overhaul at Derby Works in 1987. *(Tony Shaw)*

being used for unscheduled component changes and attention.

A change in maintenance policy came into force across the whole railway in 1987, which saw an end to the old works-based overhauls and the beginning of the new Component Exchange Maintenance (CEM) system. This was to result in selected local depots being upgraded to become 'Level 5' depots able to carry out such work. Many elements of the previous 'light' overhaul were to become the new 'E' exam, while elements of the previous 'intermediate' and 'general' overhauls became the new 'F' and 'G' exams. These had an emphasis on replacing major components (such as engines, cooler groups etc.) only when these items had reached the point when they would need exchanging to prevent the power car becoming unreliable or the component sustaining a more expensive major failure.

The new CEM system was often known as 'cost effective maintenance' as the previous system saw the major works locations stripping power cars right down to bare shells at intervals far more frequently than they actually needed for that degree of work to be carried out. Not only was that very expensive in terms of overhauling components, it also meant vehicles were out of service for far longer than necessary. It should be noted that with the benefit of experience, overhauls of the Paxman Valenta diesel engine (by far the most expensive item to overhaul on an IC125 power car) were gradually extended so that engines would receive a half-life overhaul every 18 months (approximately 6,000hrs of

CENTRE The inside of a power car stripped of its engine and cooler group as work gets underway on another overhaul.

LEFT No fewer than eight Marston cooler groups lined up at Crewe Works awaiting overhaul before installation back in power cars. Large component overhaul is carried out by specialist facilities.

RIGHT Not a job for the depot, or at least not yet. No 43137 is seen at Laira on 17 November 1986 awaiting a tow to Derby Works for collision damage repair. *(Kevin Daniel)*

CENTRE No 43082 was the first power car to receive a new cab at a depot following collision damage, this work previously being the domain of major works sites. The job is well underway at Neville Hill on 11 September 1997. *(Paul Corrie)*

operation) and a full overhaul every three years (approximately 12,000hrs of operation).

As far as IC125 power cars were concerned, two depots were chosen to receive the required expenditure in equipment and manpower to become 'Level 5' depots, these being Bath Road (Bristol) and Neville Hill (Leeds). For a transitional period between Autumn 1987 and Spring 1988 the local depots carried out their first 'Level 5' overhauls while Derby Locomotive Works was winding down, ready for closure.

In May 1988, the final overhauled power car left Derby Works leaving heavy maintenance entirely in the hands of the local depots, although trailer sets still returned to nearby Litchurch Lane for refurbishment and other attention. The entire fleet of Western-based power cars would receive their major exams at Bath Road. Some Eastern power cars would also visit Bath Road for major attention, as Neville Hill would not have the capacity to deal with the sizable Eastern and Scottish-based power car fleet on its own.

With the benefit of experience, a Paxman Valenta power unit could be changed in 24 hours (three shifts) including disconnection of the old engine, lifting it out, and lowering the new one in, reconnecting the replacement, and testing. Prior to depot-based exams it would have typically have taken a day just to transfer the power car to Derby.

Bath Road depot was closed in 1991 and the major exams were transferred to Laira. The same year, the completion of the ECML

RIGHT Neville Hill's own Nos 43075 and 43089 undergo attention on 14 January 2018.

(John Zabernik)

RIGHT Nos 43083
and 43044 have been
carefully drawn close
to the gantry for
attention at Neville Hill
on 16 September 2017.
(John Zabernik)

electrification scheme resulted in the cascade of most power cars from that route, ending the need for non-native power cars to receive their major exams on the Western Region. After privatisation, the owning leasing company became responsible for arranging the major exams on its own vehicles rather than the Train Operating Company (TOC), but with no immediate alternative available, Laira and Neville Hill continued to carry out exam levels 'E' to 'G' as they had done previously.

Evolution of train leasing agreements introduced two new terms into IC125 maintenance lexicography. 'Soggy lease' became used to describe the leases as originally set up whereby day-to-day maintenance was carried out by the operator, while heavy maintenance (exam levels 'E', 'F' and 'G' along with engine and cooler group exchanges) remained rolling stock leasing company responsibility.

'Dry lease' was the term used where the train operator took responsibility for all exams and repairs, and this has enabled them to reduce operating costs by carrying out the required work at their own depots. For power cars on the ECML route this resulted in a combined overhaul and reliability improvement programme undertaken at Doncaster Works from 2001 to 2004. Routine exams up to and including 'G' exams were carried out at its own depot at Craigentinny.

For power cars on the Midland Main Line route, Neville Hill continued to handle all levels of work, but a departure from the previous rigid exam sequence to a more responsive regime coined 'X' exam allowed component life to be extended. Power cars were subsequently treated on a more piecemeal basis rather than receiving all the required work in a single exam.

Western Region power cars continued to receive routine major exam attention at Laira, although those from the First Great Western, new measurement train and Midland Mainline fleets visited the Devonport dockyard site in Plymouth for attention by private contractor DML between 2001 and 2006. DML also carried out overhauls to return six long-term stored power cars to service for Grand Central in 2006/7. Subsequent 'E' exams on that fleet

BELOW No 43064 is
seen on jacks within
the workshop at
Neville Hill on
13 September 2009.
(John Tattersall)

RIGHT Nos 43076 and 43061 are standing in the Heavy Repair Shop at Neville Hill on 26 May 2013. Parking the power cars the same way around makes the job far easier with regard to the position of equipment. *(John Tattersall)*

CENTRE A final but important part of a power car overhaul is the repaint. Not only does the paintwork portray the right image to the passenger but it protects the bodywork from corrosion. This is No 43076 rubbed down ready for a new livery to be applied at Neville Hill on 21 February 2010. *(John Zabernik)*

were dealt with by its home depot at Heaton (Newcastle) and 'F' exams were mostly carried out by LNWR, Crewe.

Chapter Six deals with the 'repower' life extension programme undertaken principally between 2006 and 2008. For all power cars that received the MTU 16V4000 R41 engines this meant a visit to Brush (Loughborough) which began handling IC125 power cars in 2003. It continues to undertake IC125 work, having taken on most of the specialist overhauls which fall outside the remit or capability of the depots.

Derby Works had handled all accident damage repairs until its closure in 1988. Generally, such work was beyond the scope of the 'Level 5' depots as they lacked both the capacity and the experience to effect repairs without impacting on their regular workload. Doncaster Works handled some collision damage work which arose, until 1991, but after that Crewe Works received any vehicles with accident damage.

Some spare driving cabs had been produced when the power cars were constructed, but the last of these was used on No 43158 following a collision in the Severn Tunnel in December 1991. Subsequent seriously damaged power cars spent long periods out of traffic while further spare driving cabs were produced. Notably, this resulted in No 43071 and

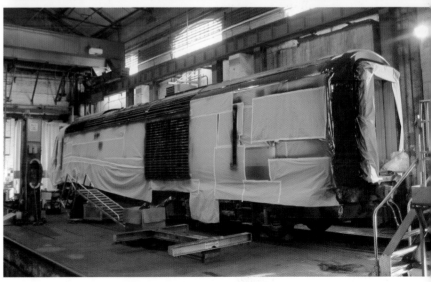

RIGHT No 43045 is shrouded in masking during application of the complex East Midlands Trains livery at Neville Hill depot on 7 November 2010. *(John Zabernik)*

No 43180 spending more than two years out of service following accidents during 1994 at Newton Abbot and near Edinburgh respectively.

Crewe Works handled a small number of other power cars returning to their place of construction for attention. In 1995/96, Crewe undertook the conversion of Nos 43167–43169 to VP185 engines, the revival of No 43104 after a long period in storage in 2001, and finally handled the overhaul of both Nos 43099 and 43102 in 2004 when transferred to Great North Eastern Railway. Crewe was also used for its ability to store very seriously damaged vehicles out of public view while accident inquiries were completed, prior to the vehicles being discreetly disposed of.

Following the tragic accidents at Southall (West London) in September 1997, Ladbroke Grove (West London) in October 1999, and Ufton Nervet (Berkshire) in November 2004, vehicles damaged beyond repair, including power cars Nos 43173, 43011 and 43019 respectively, were received by Crewe. Collision damage work, including driving cab replacement, switched to being handled by the major depots beginning in 1998, when Neville Hill undertook repairs to vehicles involved in minor accidents at the depot. Subsequently, Laira and Craigentinny have both undertaken cab replacements on damaged vehicles.

Power car exams A to G

Every time an IC125 visits a depot it is given the most basic attention. This includes refuelling and checking of the oil, coolant and windscreen washer levels, and a cursory look through and around the train checking for obvious defects. Seals on equipment that can be isolated will be checked to ensure they are intact and the defect book will be checked for new entries. Over and above these daily checks, a series of examinations are scheduled. The frequency of these exams is determined by the train mileage operated. Each exam also includes all elements which would be undertaken in the lower level of exam, *i.e.* a 'D' exam also includes everything that would be undertaken during exam levels 'A' to 'C'.

The higher level exams ('E', 'F' and 'G') are carried out approximately every two years and the cycle will be G, E, F, E, F, E, G, so the biggest 'G' exam is undertaken every 12 years. Under British Rail operation, higher level exams could be deferred or downgraded to save money, such that only the safety-related elements such as running gear would receive attention. Since privatisation, the rigid system is adhered to even if the power car is only expected to remain in service for a matter of

BELOW Craigentinny was never the best equipped depot, but they managed overhauls on a fleet of more than 40 power cars at their height. No 43320 is seen receiving an engine change using the outside crane. *(Peter Starks)*

months after the 'G' exam is completed. It should be noted that power unit and cooler group changes are undertaken when they become due outside the 'A' to 'G' exam system to maximise the life of these items.

- ■ 'A' Exam. Underframe checks (brake blocks and pads, gearbox oil-level, gearbox flexible drives and brake discs), front end lights, wheel slide prevention test, functional test of cab air con, test of luggage van doors, cooler group oil level and general checks, general exterior inspection (paint damage etc.), fire-system bottles (integrity and security).

- ■ 'B' Exam. Brake gear inspection, traction motor internals check, dampers and wheelset visual inspection, checks to underframe, battery examinations, driver/guard communications, and compressor contactor checks.

- ■ 'C' Exam. Compressor checks (efficiency, filters, governor), main reservoir pressure, air pipes visual inspection, brake system and parking brake tests, calibration of brake gauges, more thorough checks of bodysides and internals such as roof hatches, and lubricate, gangway examination, nose-end lights examine and clean, AVR examine and battery voltage check, engine coolant level switch check, internal lighting clean and check, traction motor cabling insulation test, wheel slide prevention equipment checks, cab floors seat and internal fittings check, cab air conditioning performance check, traction motor blower air flow test, fuel lift pump brush length check, lubrication of traction motor armature bearings, luggage van sliding door lubricate and door lock mechanism checks, traction motor contactors, reverser & PEFR contacts/circuits test, cooler group clean and air flow test, AWS/TPWS and DSD test, traction motor gearbox reaction links and support brackets check, traction motor gearbox lateral movement check, emergency couplings and Alliance buckeye checks, bogie tread hangers check, bogies and underframe clean, brake disc distortion checks, wheelset measurements and fire system test.

- ■ 'D' Exam. Compressor oil change, automatic drain valve test, air system safety valve test, compressor unloader examine, parking brake exhaust valve, manual release and pressures check, exhaust deflector and luggage van roof vents examine, cab windows examine, cab and van luggage rails examine, battery overhaul, 36-way and ETS jumpers and receptacles examine, ETS interlock test, fault indication lamps test, starter motor contactor box check, traction motor junction box clean/examine, ETS cabling insulation test, floor grating examine, compressor motor bearings lubricate, alternator rotor bearing lubricate, main alternator cables examine, luggage van doors lubricate, short circuiter examine, cooler group hydraulic oil and filter change, traction motor gearbox oil change, fire-system hoses check, rescue and windscreen ladder condition check, overhead live wire warning signs clean.

- ■ 'E' Exam. Bogie change, compressor change, headstock air hoses examine, air system drain tank examine, air system safety valve test, windscreen wiper equipment overhaul, main reservoir pressure check, air brake pipe gauge calibrate, body to bogie flexible hoses renew, air and brake system full test, bodyside panelling examine, headlamp panel clean and check, cab door drop window examine, power car interior clean, gangway examine, cab windscreen examine, battery change, monitor card battery renew, battery charger check, contactors resistors and spare fuses check and examine, engine coolant level sensor clean and calibrate, 36-way and ETS receptacle examine, cab gauge illumination check, general fault indication check, speedometer test, relays examine, control cubicle examine, rotary switches examine, traction motor blower air flow test, internal and external doors examine and lubricate, traction motor contactors examine, reverser change, cooler group hydraulic oil and filters change, cooler group performance check, DSD pedal unit change, emergency coupling and nose-end door examine and lubricate, coupler and drawgear overhaul, and drawbars examine.

- ■ 'F' Exam. Warning horns renew, air reservoir tank straps examine, safety valve change, warning horn manifold renew, brake controller change, emergency brake valve change, E70 change, EP brake valves change, pressure control valve change, parking brake pressure reducing valve change, double and duplex check valves change, cab and cab side valances check, removable roof panels clean, gangway overhaul, starter motor contactors change, alternator exciter contactor change, engine coolant level sensor renew, cable ducts and terminal boxes examine, rectifier RFR fault relays test, pushbutton switches examine, fuel lift pump change, fuel and lubricating pipework examine, fuel tank hose renew, fuel, cooling system and lubricating oil self-sealing couplings change, traction motor blower clean and examine, cubicle fan renew, traction motor contactor change, main rectifier examine, AWS/TPWS reset switch and electronic alarm/indicator unit change, AWS/TPWS receiver junction box check and clean, emergency drawbar change, battery box repair, removable roof panels paint, engine room paint, roof paint, underframe touch in paint, cooler group bay paint, and body exterior repaint.

■ 'G' Exam. Air compressor silencer change, fuel tanks change, compressor expansion chamber examine, antifreeze unit change, air reservoirs change, air system drain cocks examine, cab and air system flexible hoses examine, miniature circuit breaker panel date check, master controller examine, power cable insulation test, surge suppression equipment change, and emergency drawbar examine.

MAINTAINING A SINGLE POWER CAR

An IC125 power car will receive an overhaul after between 488,000 and 600,000 miles, depending on the operator. This equates to around two to three years of running. When the fleet was first introduced, overhauls were carried out every year. The period between has been pushed out steadily since then as experience has been gained, without compromising safe operation. There are three levels of overhaul ('E', 'F', and 'G' exam), each progressively more extensive. An 'F' exam, for example, will cost over £200,000.

The journey to reliability

Although the prototype set had accumulated considerable mileage its use was nowhere near as intensive as the production sets were to encounter, and reliability problems soon manifested themselves. Some of these were design flaws which had not become apparent until in-service experience had been gained, whilst others were the result of changes to the design or specification of materials used on the IC125.

One of the earliest significant problems arose when defects were discovered on the BT10 bogies used on all the trailer vehicles. The problem was first highlighted by a potentially disastrous fracture on the primary suspension arm of a locomotive-hauled Mk3 during mid-1976. This resulted in urgent checks of the identical IC125 trailer bogies. Cracks were found on so many bogies that for a few days test running and driver training of the new trains was halted and it looked likely the launch of the 125 service would need to be delayed. Closer examination of the bogies indicated most of the cracks were not critical and they could continue operating, provided they were closely monitored.

A small team of senior technical staff was established at both Western Region depots tasked with examining the axlebox spring castings. A procedure was devised which allowed them to be checked with the wheelsets still in situ. Preliminary work had identified the risk area as the rear face of the casting and each seven-carriage set had 56 such castings, which needed to be examined. The very limited clearance between the casting and the wheel called for an innovative examination technique involving the use of a penetrant magnetic dye. After cleaning the area to be tested it was sprayed with a pink dye, a strong magnet attached and then the area was sprayed with a solution of iron filings which would line up with any cracks present at the surface.

Visual examination involved shining a torch up from below while holding a shaving mirror at an angle over the top of the bogie frame. Some vehicles were so bad they required immediate attention, and these received a prominently painted red cross on the casting, whilst others considered to be worthy of monitoring for attention at the next opportunity, were marked with a blue cross. Initially, defective castings were replaced by those of the same type but by the end of 1976 a re-design to a much heavier, better machined type allowed a fleet-wide replacement of all castings to commence, beginning with the most obvious candidates.

It was very fortunate that no castings failed in service at speed as a derailment would have been almost inevitable. Another early bogie problem was found on the power car yaw dampers. The bolts securing them in place could not cope with the forces applied so new, stronger bolts had to be retrofitted fleet-wide. Pins holding the traction motor gearbox in position on the bogie were to be the source of another early urgent safety check. One failed in service with the potential to allow the gearbox to drop onto the track, so securing pins were checked and gradually replaced.

Despite having accumulated a lot of experience with the Paxman Valenta engine while operating the prototype, a casting problem was found with the piston on an early production power car following a failure during test running. It was found that significant cracks were emanating from the piston gudgeon pin hole and these caused oil to be passed into the exhaust. A new design of piston was retrofitted to the engines already supplied, by the depots in Bristol and London, rather than hold up the new build production line by waiting for Paxman to swap the whole power unit over.

ABOVE Not for public consumption – or at least not yet! The launch of the new Midland Mainline livery took place on 10 February 1997. In readiness for this, Nos 43059 and 43058 were repainted at Neville Hill prior to running light to Derby under the cover of darkness the day before, to join a matching rake of carriages for the high-profile launch at London St Pancras. (Paul Corrie)

BELOW No 43101 *The Irish Mail/ Tren Post Gwyddelig* passes Didcot North Junction while working the 11.06 Newcastle to Bournemouth on 31 July 1999. Three Virgin Trains power cars received names applied in the form of sizable vinyl decals during Summer 1998, before the company decided to return to cast nameplates for subsequent dedications. Mixed-livery formations were commonplace at this time when vehicles were gradually repainted into the new livery schemes. (Martin Loader)

RIGHT **The 11.00 London Paddington to Bristol Temple Meads First Great Western service (headcode 1C11) is kicking up so much snow at Shrivenham on 7 January 2010 that the rear of the train is completely blotted out. No 43154 leads the train past the site of Ashbury Crossing.**
(Martin Loader)

BELOW **Grand Central's No 43468 is seen running 'light engine' over Skelton Bridge Junction while returning from Crewe LNWR to Heaton depot on 24 October 2014, following maintenance attention.**
(James Trebinski)

The problem of engine exhaust deposits fouling power car cabs became evident as soon as trial running commenced. With the power car at the rear of the train the slipstream would force exhaust over the power car roof and windscreen making the trains look dirty. Also, the exhaust contained oil which made windscreens difficult to keep clean. Cleaning the windscreen safely was a particular problem under the overhead wires at King's Cross. In order to overcome this, an exhaust deflector was designed which took the form of an additional sheet of metal attached to the roof above the clean air compartment from the cab to rearward of the exhaust portals. A gap between the power car roof and the exhaust deflector induced a layer of clean air between the cab and the exhaust flow and proved effective in significantly reducing the problem. Nos 43012 and 43013 received the deflectors experimentally from September 1977 and fleet-wide modification began in 1978 with the last four power cars built in the main ECML batch (Nos 43118/119/122/123) receiving them from new. The rest of the ECML power cars were retrofitted with deflectors at Stratford (East

London) depot prior to service introduction. Retrospective fitting of deflectors on the Western Region was erratic with power cars sometimes missing fitment during a works visit, so it was not until April 1981 that the last one (No 43052) received the modification.

In later years concern about the exhaust deflectors cracking around the exhaust portals, resulted in a modification to those working on the East Coast Main Line from 2001, the modified deflectors having extra material added around the edges. In more recent years the introduction of the much cleaner MTU engine has negated the need for the exhaust deflectors so they have been removed from the fleets maintained by Craigentinny. On the power cars operated by Great Western the exhaust deflectors have been shortened so they omit the section around the exhaust portals, but the rest of the deflector remains in place as it was decided they also protect anyone working in the clean air compartment in the unlikely event of an object being dropped on the train and penetrating the fibreglass roof section.

An unforeseen early problem encountered was with the control system. The interlock with the traction motor contactors and the reverser switch was getting stuck in position. This became apparent after the train changed direction at a terminal station; the faulty power car would try operating in the original direction meaning that rather than working together the power cars were effectively working against each other. The alternator in the defective power car would not excite, but the traction motors would generate an electric braking effect, as the healthy power car started to pull away. This braking effect would counteract the power produced and bring the train almost to a stand still at around 5mph. A re-design of the spindle in the interlock control was implemented which cured this issue.

A couple of problems arose from changes in design between the prototype and the production sets. The traction motors were bogie mounted to reduce the unsprung mass but the axles are driven through a flexible drive and gearbox. The flexible drive is essential to allow the axle to move up and down relative to the fixed traction motor. To improve lubrication of the gearbox bearings and extend the period between servicing a new cast light alloy gear case with a larger oil capacity replaced the cast iron unit on the prototype. In service experience showed the gears were not receiving adequate lubrication, oil was leaking through the seals, and once the oil level had fallen sufficiently gearboxes could seize. In some alarming cases this resulted in a locked axle leading to a wheel flat as the axle was dragged along the track. Ultimately a wheel set with a seized gearbox was to result in the serious derailment of the 13.00 London King's Cross to Edinburgh service near Northallerton on 28 August 1979. Thankfully, nobody was seriously injured in that derailment and the train (ECML set No 254028) remained upright, but a solution was needed. A new design of oil seal was introduced; the root cause of the failures having been found to be air passing through the seal when the train was running at speed, which pressurised the gearbox forcing oil out. The gearbox oil fillers were also moved, a sight glass was added allowing them to be checked on the 'A' exams and the oil capacity was increased allowing a train with a failed seal to operate safely until it could be attended to. Failure of gearbox oil seals is still an issue today. The most easily visible clue to this problem is a streak of oil on the bottom of the power car bodyside emanating from the bogie.

The Northallerton derailment resulted in a short-term modification to the wheel slide protection equipment: a seized gearbox would illuminate the wheelslip warning light whenever the train was running above 6mph. Reliability of wheel slide protection was such that numerous spurious indications were received so it was further modified so the wheelslip protection system detects seizure under power and the wheel slide protection takes over detection when coasting or braking. Power cars in some fleets were eventually fitted with a 'Locked Axle Indicator' on the cab desk. If this illuminates red the driver must stop immediately and check the wheels are rotating freely.

A design change on the turbocharger turbine inlet nozzles came about as the foundry which supplied the nozzle rings to Napier for the SA-084 turbocharger closed down after completion of the first batch of production power cars. Unable to source another supplier

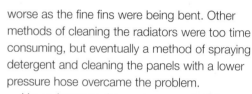

to produce the original 32-vane nozzle ring resulted in a 24-vane unit having to be used instead. These were failing at a significant rate, with cracks found between the ring and the nozzle blades on failed units. Although a design change and a higher strength alloy was tried the eventual solution was to source another supplier to manufacture the original 32-vane design.

Bolts securing the cast iron brake discs on the power car wheelsets were to be another cause of failure as the amount of heat dissipated during braking was causing the bolts to fracture, or the disc itself to fracture at the bolt positions. A faster method of extracting the remains of the failed bolt from the wheelset was devised to speed up replacement of broken bolts. To counteract the cracking discs an improved, self-venting disc with radial fins on the inner face in contact with the wheelset was developed. This allowed heat to dissipate outwards from behind the disc during braking.

With seven years of in-service experience, IC125 engineers had largely overcome the early problems but an unusually hot summer in 1983 resulted in a record number of engines reverting to idle or shutting down entirely. Two significant but related problems were found: inefficient cooler groups and cracked exhaust manifolds. Radiator panels were becoming blocked externally by dust, dirt and seeds etc, being sucked into the cooler group and then sticking to surfaces coated in oil expelled by the fan drive. Early attempts to clean the radiators using high-pressure hoses made the situation

worse as the fine fins were being bent. Other methods of cleaning the radiators were too time consuming, but eventually a method of spraying detergent and cleaning the panels with a lower pressure hose overcame the problem.

Hoses between the cooler group and the radiator were found to be collapsing inward with age, which was reducing coolant flow. This was being exacerbated by jointing compound in the engine assembly dissolving in the hot coolant and settling in the radiator panels reducing their internal efficiency. Cracks were appearing on the original aluminium exhaust manifolds (specifically at the turbine inlet bend) and on the engine crankcase. An improved aluminium exhaust manifold was eventually produced using stronger alloy which reduced the failure rate of this component down to virtually zero.

The inefficiency of the cooler groups was part of the reason for these failures and once cracks appeared, coolant was able to escape which terminally sealed the fate of the system, but a previously underestimated phenomenon was being studied and understood. When full power is selected the engine heats up and all its components expand, once power is shut off again the engine cools down and contracts. This is known as thermal cycling. In other diesel engine applications the engine will remain in full power for a sustained period and return to idle less frequently. On an IC125, particularly one operating on a start-stop service, monitoring equipment found demand for power was frequently changing. On average the Paxman

Valentas were going through ten thermal cycles an hour for 15 hours a day. This was putting enormous stress on engines, which had not been foreseen and failure of the engine at its weakest point was inevitable.

The turbine inlet bend was redesigned with an improved component which reduced the failures down to a negligible level, but the cracks in the crankcase were to be more of a problem. When engines were first delivered they were found to be overweight, and this was cured by machining metal off the top deck of the block, resulting in a recessed deck. The reduction in thickness created a weak point where the cracks and coolant leaks initiated around the top of the cylinder liners. Later batches of engines tolerated the excess weight rather than weaken the crankcase, but the remaining recessed deck Valenta engines continued to be prone to coolant leaks until the type was phased out in 2008.

To try to overcome the coolant issues, two different approaches were taken. On the Western Region, additional coolant bowsers were located on station platforms to top-up the system in service. On the Eastern Region

(where typically, trains ran much further between stops) an additional coolant tank was installed in the luggage van from 1985, which effectively expanded the capacity of the header tank. This equipment included an additional filler point, an external drain tap located on the bodyside grille beside the cooler group, and an old air tank re-purposed to contain coolant. This equipment was removed when it was found the additional

BELOW No 43307 is seen leading the 17.33 King's Cross to Harrogate past Claypole on 5 May 2017. Dirt spoiling the bright red Virgin Trains East Coast livery above the leading bogie shows this power car is suffering from leaking gearbox oil seals. This is now perfectly manageable following modifications made to the fleet. *(Martyn Spencer)*

ABOVE An exhaust deflector similar to that fitted to all power cars, but in later years only retained by the 24 East Midlands Trains vehicles.

ABOVE RIGHT A close-up of the modified exhaust deflector from an East Coast power car, with strengthened edges and exhaust port surround.

pipework and connections resulted in more leaks and the only evidence remaining is a gap in the external radiator grille on certain power cars.

From October 1985, failures to the exhaust manifolds started to increase and by February 1986 the failure rate was greater than the earlier replacement rate prior to the improved aluminium manifold being introduced in the light of the 1983 difficulties. By June, manifolds cast in late 1985 and early 1986 were also failing and this confirmed it was not simply a case of the replacement units reaching the end of their fatigue life. British Rail and Paxman engineers worked all summer to determine the cause of the failures with Paxman initially dumbfounded at the scale of the problem and at a loss to explain its sudden onset.

Static testing of power units failed to replicate the fault and studies were carried out on the failed exhaust manifolds as it was felt that a manufacturing defect was the only plausible cause of the problem. It was then found that the onset of the troubles coincided with the introduction of a new type of antifreeze across British Rail. The new antifreeze solution was attacking the surface of the aluminium with a chemical designed to prevent corrosion of ferrous metal components, which deliberately corroded the surface of the metal with the resultant layer acting as a barrier against further attack. Tests conducted at the GEC laboratories proved that the surface of the manifolds developed cracks when exposed to the new antifreeze, which in service would soon

became deeper due to the heat and stress placed on the component during operation. Operational experience and study of failed units showed that once the cracks in the exhaust manifold had become deep enough the component failed, resulting in sudden and sometimes spectacular coolant loss causing the power car to shut down. The previous plans for switching to cast iron exhaust manifolds were quickly reinstated and soon production started on the new, heavier spheroidal graphite cast iron manifolds, and these were fitted to power cars to replace failed units as the summer progressed. Once the antifreeze problem had been discovered it was quickly rectified and Shellsafe antifreeze reinstated across the whole of British Rail. The combination of the new cast iron manifolds and the replacement antifreeze solution consigned the problem to history, but not before an expensive and embarrassing lesson was learnt.

The stress of the IC125 work environment inevitably put enormous pressure on the Valenta engines. In Chapter Six the various replacement engines are detailed, but for the first 30 years of operation the original power units remained in service across the vast majority of the fleet. One of the most frequent faults on the engine itself was caused by the failure of the piston rings owing to wear and tear. If the piston ring failed the combustion gases could bypass the ring and pressurise the engine. This would force engine oil out of the roof-mounted oil breather, sometimes resulting in the entire 'A' side of the power car becoming covered in oil. When compared with the MTU R41 engine, fires involving the Valenta were relatively rare, but when they did occur they were often the result of fuel passing into the turbocharger or the turbocharger itself seizing up. Whilst such fires would look spectacular they would quickly burn out with little damage to the power car. Water-cooled exhausts on the Valenta (and the later VP185) have often prevented or mitigated fires which the power car's own in-built fire system have failed to contain.

In the face of falling subsidies, cost reductions were sought and in May 1988 British Rail invited tenders for the replacement of the entire IC125 engine fleet. They received four submissions, but decided that replacement of the now sufficiently reliable Paxman Valenta would not be cost effective. In 2001, 15 of the recessed-deck Valenta engines were modified by extra material being added to strengthen the deck, bringing those engines up to the standard of their flat-deck counterparts, thus extending the life of those engines and reducing the failure rate from coolant losses. This work was carried out by ADtranz and the modified engines were given numbers prefixed 'AD' to distinguish them from unmodified engines. Similar work was carried out on additional engines during subsequent overhauls, further reducing the number of recessed-deck engines in use. Several of these were renumbered with 'TD' prefixes to denote 'Thick Deck'.

More than half the Valenta fleet was modified, beginning in 2003, to provide Single Bank Firing (SBF) while idling for a period of time. This enabled the engine to run on six of its twelve cylinders, reducing fuel consumption and exhaust emissions. The system was controlled by an SBF unit located on the 'B' side wall of the clean air compartment, which operated a collapsible linkage on the fuel rack of 'B' bank side.

Ten additional Paxman Valenta engines were purchased in 2004 to increase the available pool of engines, mitigating the loss of some engines from the original 233 supplied between 1975 and 1981, which had suffered from terminal, non-repairable failures over the years. Remarkably, these additional engines were of the recessed-deck design.

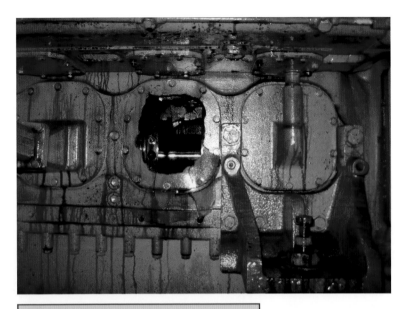

FILLING THE TANKS

On average, an IC125 power car will use around 3.6 litres of fuel every mile. So, a London to Edinburgh journey will consume 2,830 litres of fuel (£3,600 at 2018 pump prices) and London to Cardiff cost £1,325 (1,044 litres). Each IC125 power car holds 4,680 litres of diesel. At a typical UK 2018 pump price of 127p/litre, that is £11,887 to fill both tanks from empty!

ABOVE A piston has smashed through the crankcase on the 'A' side of engine No AD7 while installed in No 43123. An expensive (if not terminal) event for the engine.

BELOW No 43014's engine has forced much of the contents of the sump out through the oil breather, turning the Network Rail yellow-liveried power car black. This view, dated 16 September 2003 also shows the coolant tap on the radiator external grille from the expanded header tank modification. *(Tony Shaw)*

Chapter Four

In the driver's seat

Many little boys (and a some girls!) dream of driving the mighty Inter-City 125 at full speed. This chapter looks at what has become a routine task for thousands of train drivers all over the country since 1976, including some who still secretly get a child-like thrill from taking the controls of the iconic train.

OPPOSITE No 43190 approaches Circourt Bridge, Denchworth on 9 October 2003 with the 07.30 Swansea to London Paddington. This livery was introduced in 1999 after First Group became the sole owners of the Great Western franchise. It was applied using vinyl over the earlier green and ivory scheme. *(Martin Loader)*

LEFT The transition from the original blue and grey livery to the new 'Executive' colours is seen here on 24 February 1986. No 43143 arrives at Reading leading the 14.30 Paddington to Bristol Temple Meads with uniform-liveried set No. 253 037, and power car No 43016 at the rear. It is passing a set in the original colours, departing for Paddington. There were two drivers in this era on all services scheduled to operate at 125mph. *(Kevin Daniel)*

CENTRE The coastal section along the sea wall between Dawlish Warren and Teignmouth in Devon is one of the most famous railway locations, and has been traversed by IC125 sets for nearly 40 years. On 22 June 1996, the 07.35 London Paddington to Penzance nears Teignmouth, powered by Nos 43171 and 43130 *Sulis Minerva*. *(Chris Hopkins)*

In years gone by the career progression of rail staff to driving top link traction such as the IC125 was generally a long, drawn-out process. Until 1988, staff worked their way up into the 'footplate' grades (*i.e.* those employed to work in the driving cabs) with all drivers starting out as Secondmen or Drivers' Assistants. They gradually learnt the job from experienced hands before progressing into the role of driver when a vacancy arose. Once 'passed out' as a qualified driver they would be placed in the most junior link, generally just confined to localised shunting work or driving shorter distance local work. As more senior drivers retired or left the profession, drivers would work their way up through the links (rota positions) which meant the next most senior driver would move up a position and receive any additional training their new link position entailed.

Much changed between 1988 and 1998 when a new process was devised to match

LEFT No 43135 passes Circourt Bridge, Denchworth on 21 November 1996 with the 12.15 London Paddington to Bristol Temple Meads Great Western Trains service. A complete train in the then new livery was not that common at the time. Any livery incorporating this much white does not tend to stay pristine for long. *(Martin Loader)*

modern training and recruitment needs, and new drivers would now undertake a year-long course combining classroom and practical experience before being assessed and passed out as a qualified driver. Under British Rail, trains scheduled to run at over 100mph still required a second, fully qualified driver in the cab. This included most IC125-operated services on the Great Western and East Coast Main lines. Drivers had received a comparatively low salary, but their income was supplemented by a series of bonus payments, one of these being based on distance covered (known as 'mileage'), which further cemented the position that the IC125s would generally be driven by senior drivers.

The restructuring of drivers following privatisation swept all this away. In return for a higher salary the trains would be driven by a solo driver at speeds up to 125mph and the various bonus payments were incorporated into the salary. Another two factors affected driver recruitment following privatisation. First, many of the Train Operating Companies (TOCs) were seeking to run more services, hence needed more drivers and secondly, some of the longer distance Inter-City operators had inherited a higher ratio of drivers approaching retirement age when larger depots were split up, with drivers divided into one of the new TOCs or freight operators. This chapter recounts the reminiscences of a driver working for First Great Western from 2002 until 2006.

I joined the railway in April 2000, working as a member of station staff, but by September I'd started on a driver-training course working for a division of the old Regional Railways on local train services in the Bristol area. As was common by this time, most of the other trainees on the course with me were 'straight off the streets' (*i.e.* they had been recruited externally directly into the role of train driver). By April 2002 I'd moved onwards and upwards and was starting with First Great Western based at Bristol. The leap in salary was attractive enough in itself, but driving IC125s at full speed on the Bristol to London route was soon to become a reality. The initial training period was an interesting time in itself. Anyone and everyone involved in training at Bristol was busy training drivers on the new Class 180 diesel multiple units so myself and the other new

ABOVE A typical driver's eye view. This was taken from a GNER power car while stationary at a red signal outside Leeds as another IC125 departs.

recruits were dispatched to Plymouth for three weeks to learn about the IC125s.

Our instructor was a very friendly chap with lots of experience and enthusiasm who looked after us superbly – as well as enjoying three weeks of living in a hotel with everything paid for by the company, being at Plymouth brought other advantages. A small room at Laira depot was commandeered as our classroom, this

BELOW No 43164 pauses at Swindon while working the 09.40 Bristol Temple Meads to Paddington on 15 September 2003. Trains on this route made up the bulk of the work undertaken by Bristol-based drivers. This livery was nicknamed 'fag-pack' as it resembled a contemporary cigarette packaging. It was introduced in 1999, but largely eradicated by the end of 2003. *(Gary Heelas)*

was just a short walk to the main sheds of the company's biggest IC125 depot so we could quickly become hands-on by looking at an actual power car if learning from a book became too tedious or challenging. As Laira was host to power cars in various stages of heavy overhaul we seized a few opportunities (with the permission of the maintenance staff) to go into, above, around and underneath power cars with major components removed allowing us to see bits of them few trainees or indeed few drivers will ever have seen.

At the end of the first week with much of the classroom activity out of the way we were unleashed (albeit under *very* close supervision) on our first actual train. My colleague drove from Plymouth to Newton Abbot and I then took over the controls of 43183 leading 1A56, the 12.38 Plymouth to Paddington from Newton Abbot to Exeter along the famous coastal line. The following week we undertook a few runs on the Plymouth to Exeter section which was ideal for getting a feel for the train. Between Plymouth and Newton Abbot the route is mostly 60mph and includes some steep gradients so a driver is always doing something, either applying power or using the brake. Between Newton Abbot and Exeter the route is much flatter and a short stretch of 100mph gives a chance to experience the train at speed.

We were taught that when starting away we should release the brake whilst selecting notch 1 on the power controller simultaneously. Once the brake was released, power should be opened up to notch 3 and the ammeter observed rising and settling in the higher power position before then selecting notch 4. Once the power car had settled in notch 4 then full power (notch 5) could be selected, this would generally be at around 12-15mph. Selection of full power from idle should be avoided; this could result in a jerky start and sometimes overload the power car. In any case, application of power would not be any better as the engine governor is still catching up with the change in power controller position. Opening up the power controller any more quickly is pointlessly trying to beat the power curve of the 2,250hp engines. When shutting off power, the power controller should be returned to notch 1 and again the ammeter observed as falling back and settling with minimal power

applied before shutting off the power controller back to idle. This was to allow engine speed to fall and avoided damage to electrical systems. During normal braking step 1 (initial) on the brake controller should be selected, once the brake has taken effect in step 1 then step 3 or 4 should be used for speed reduction of the train until reducing the brake application or returning to the running position as appropriate. Although it varied from train to train, the brake propagation (delay in releasing the brake) equated to around 5mph per step of brake applied, so for example, if aiming to reduce train speed down to 50mph a step 3 brake application should be released entirely at around 65mph. The time taken for the whole train braking system to react to the releasing of the brake from step 3 would reduce train speed by another 15mph so the new target speed would be hit perfectly. When coming to a stop the driver should aim to have reduced the brake application to step 1 by the time the train is stationary and release the brake momentarily just before the train comes to a stop to effect the smoothest possible stop. After coming to a standstill, the brake should be moved back into step 1 to hold the train stationary. This was the way experienced drivers down in Plymouth taught us to drive the trains; the methods taught at Bristol mandated a far less professional approach.

When starting away from a stand they were comparatively timid, application of power beyond notch 2 would be far more gradual and it would be far longer before the dizzy heights of full power would be reached, perhaps not until the whole train is clear of the station and a speed nearer 25-30mph attained. Braking 'Bristol style' was also more cautious, the approved style was to begin braking far sooner with a more gentle brake application, and the complete release of the brake as the train came to a standstill was frowned upon in certain quarters.

As our three weeks of training in Plymouth progressed we ventured as far as Reading three times. This wasn't strictly necessary, but the instructor had family there so our longer runs were part training, part social call! On one of the trips we needed to go via Bristol so that we could both experience driving using the Automatic Train Protection (ATP) system that was fitted to the 125mph lines on the Western Region. We'd missed our intended

train as some all-important paperwork needed completion in the office at Bristol, so instead we caught one which took the less common route from Bristol Temple Meads to Paddington via Bristol Parkway.

As our Plymouth-based instructor wasn't conversant with the route between Bristol and Swindon via 'Parkway' we were simply observers in the cab as far as Swindon, which turned out to be quite eventful. Once we got through the restrictive 25mph track layout the somewhat impatient driver tried accelerating by opening the power handle straight from idle to notch 5. The power car responded by failing to get power interlock and the ammeter dropped straight back to zero. Undeterred, the driver tried reapplying power equally as quickly another two times with the same result each time before trying a more controlled acceleration!

After being sceptical about the training benefit of us simply observing (rather than driving) as far as Swindon our instructor later revealed he was quite glad that we had seen for ourselves a perfect demonstration of how *not* to drive this train! On one of our days at Laira we gained valuable extra experience of slow speed movements around the depot by allowing the scheduled depot driver to take a break while we carried out the necessary shunts with our instructor, and another seemingly golden learning opportunity arose.

We went to move a pair of Virgin Trains power cars around the depot but the front power car wouldn't start. In a perverse sort of way our instructor was pleased as he foresaw an opportunity for us to put into practice everything we had learnt about solving

faults and failures on the power cars. He was thwarted; as soon as we opened the engine room door the low-water light on the cubicle was seen to be illuminated red, indicating the fault. Upon checking the coolant header tank gauge in the engine room we found it on zero and opening the door to the cooler group revealed green coolant gushing out of the radiator! Fault located, confirmed and diagnosed in less than 10sec – not to be outdone we had an hour of 'what could you have checked if it hadn't been that' theory while maintenance staff came and examined the errant power car before re-directing it to the shed for repair.

Once our three-week stint in Plymouth was over we spent three weeks with a Bristol-based instructor gaining more high-speed practical handling experience before being assessed; well one of us did. In what became a bit of a running joke my colleague was increasingly unwell. We diagnosed this as 'too much fresh air for a city boy' so I had the instructor all to myself for the first week enabling me to appear far more adept at handling the train when compared to my colleague during our joint assessment!

Several weeks of route learning followed, even the 'starter' link at Bristol required route learning to London Paddington via both Bath and Bristol Parkway, Bristol to Taunton (which I already knew), Bristol to Cheltenham, Gloucester to Swindon and the depots at both Bristol (St Philip's Marsh) and London (Old Oak Common). Route learning at First Great Western was very thorough and assessments were quite detailed; a written exam as well as a practical driving assessment was undertaken for each route. Learners were supported by a

ABOVE Summer Saturdays created extra demand for holidaymakers heading to/from resorts in Devon and Cornwall. To cope with this, trains not required on London routes would be used to provide the required capacity. Prior to privatisation these did not stand out too much, but the sight of a full Midland Mainline-liveried IC125 passing Middleway, shortly after leaving Par to head up the Newquay branch, may have turned a few heads. This is the 06.05 Leeds to Newquay on 20 June 1998, powered by Nos 43073 and 43055.
(Paul Corrie)

RIGHT No 43043 is seen on a misty winter morning passing through Loughborough while working the 08.05 London St Pancras to Nottingham, on 19 January 2017. (Chris Hopkins)

route-learning school which was well equipped with maps and videos of each route. The latter were particularly useful as observing unusual movements and less frequently used junctions and cross-overs at some locations would be practically impossible to observe while travelling through the area as most trains would do at full speed. The first time I drove an IC125 on my own was on 16 August 2002 when I worked 1A10, the 06.35 Bristol Temple Meads to Paddington. No 43187 led the train with 43164 at the rear – the journey was entirely unremarkable, which is exactly what any 'rookie' driver wants!

Taking an Inter-City 125 up to its maximum speed and keeping it there is a relatively simple task on the generally level high-speed lines where 125mph is permitted. Once the train has been started away from the station stop and is clear of speed restrictions sometimes prevalent on the complex track layouts around the station areas, it's simply a case of leaving the train in full power until it reaches line speed. For most sets and on most routes 125mph can then be maintained by remaining in power notch 3 with the occasional need to reduce or increase by one notch as conditions dictate. The actual skill and technique is needed when the train needs to be stopped. The distance required to stop an IC125 from full speed is between one and a quarter miles and one and a half miles. The train can stop in a shorter distance, but heavy braking would result in a rough journey for the passenger, the risk of over shooting the intended stopping point, and the ire of traction inspectors and other management! The way that drivers learn the routes leans heavily towards braking points; many drivers will use a similar 'landmark' as they approach a station as a point at which they will begin braking. The 'landmark' can be anything, but will typically be something fixed and adjacent to the railway for example a bridge, a signal, a lineside sign or a track feature. As the speed reduces, the driver may well have a further target in mind, such as (for example) the intention to reduce down to 90mph before a second 'landmark' is reached. This allows the brake application to be increased or reduced if the train brake doesn't respond exactly as expected. For the experienced, confident driver, hitting the start of the platform at around 50mph (where the line speed and signalling permits) is a target, but it's essential to state that such a speed a mere train length away from the intended stopping point is only possible if the brake application is already established in a high enough brake step (such as step three). Reaching the start of a platform at 50mph with the brake released and THEN applying it will certainly result in most, if not all, of the train stopping beyond the station! A skilled/experienced driver will actually make very few movements of the brake controller as they bring the train down from full speed to a standstill. After a brief, 'initial' application a driver will typically place the brake controller into step 3 and leave it there, only moving the controller occasionally to make adjustments in reaction to the train brakes being slightly better or worse than average.

When speed has reduced to around 20-15mph a short distance before actually stopping, the driver will release the brake or reduce it to the 'initial' position as the distance to the stopping point dictates. This results in a smoother final stop for the passengers. Drivers will use their skill and knowledge of the route to adjust the braking point in wet weather or when leaf-fall creates the risk of a contaminated railhead. If the train encounters wheel slide the braking distance will be extended and if the rails are covered in the black residue of leaves crushed by the passage of previous trains the stopping distance can be extended significantly, even with a comparatively long and heavy train such as an IC125. Knowing where the braking point is and being able to locate it in all conditions including at night or in reduced visibility is vital for drivers. The ramifications of the braking point unexpectedly moving or changing such that it gets missed (e.g. 'the white house painted black') have made interesting reading out of a few traincrew report forms down the years!

Despite the sometimes fearsome amount of noise made by the original engines when observed leaving a station or running at speed, the driving cab environment was actually quite a pleasant place to work. The noise from the massive turbocharger is above and behind you so although audible in the cab the noise isn't intrusive, in fact, more of the engine noise can be heard from the cab without the scream of the turbocharger downing it all out. The biggest sources of noise within the cabs were generally wind noise at speed from poorly adjusted, warped or ill-fitting cab doors and their rubber seals, in addition to track noise from the bogie directly beneath the driver. Having previously worked in the much more confined environs of 'Sprinter'-type driving cabs I found the cabs to be roomy and airy with a superb view of the line ahead. Cab noise became uncomfortable if driving with either or both cab doors wide open. It was sometimes necessary to drive for a while with the full effects of natural ventilation (especially if you'd taken the train over after a heavy smoker had been driving it – this was before the smoking ban), but you'd soon find yourself reaching across to slam the door shut once the speed had built up.

Another factor that sometimes tested the skill of drivers while driving with the original Paxman Valentas, with their sometimes poorly adjusted engine governors, was the phenomenon of the so-called 'super-notcher'. On some power cars movement of the power controller would result in an almost instantaneous reaction from the engine, the governor would send the engine speed far too rapidly into (and often way beyond) the desired power notch position. In perfect rail conditions this wasn't a problem, in fact, the tendency of the engine to react swiftly could be seen as a bonus, especially for a driver on their way home! What particularly interested me was learning how to control one of the 'super-notcher' power cars on a wet rail. A rapid rise in engine speed would almost certainly result in that power car encountering wheel slip and the power control system removing power only to reapply it just as rapidly, inducing a further wheel slip. While I was learning a new route I thought I'd encountered a golden opportunity: a filthy wet, horrible day and a power car I knew to be a 'super-notcher' pulled in with a driver I was sure had been driving them since they were introduced; he certainly looked old enough! On that day I took my focus off the route as I was fascinated to see how he would use his experience to tame the erratic engine governor and accelerate the train away from the station stops without slipping. I learnt little that afternoon. As the power car slipped away he tried to ignore it and unsuccessfully apply more power, then he tried to beat the system by throwing the power right off and reapplying it again immediately. Subsequent experience taught me that it's best to let the slipping power car 'do its thing' in the hope that the one at the other end is applying power normally, if the wheel slip becomes bad or sustained then reduce power by just one notch and hope it finds its feet and let things settle down before trying to increase power by one notch at a time.

By absolute contrast, the handful of VP185 engines in our fleet suffered the opposite problem: their 'power curve' meant they took an absolute age to build up speed, especially from idle. With the usual one VP185, one Valenta combination it was best to just 'ignore' the presence of the slow accelerating 'VP' and drive normally. If you were unlucky enough to get a pair of VP185s the only method seemed

ABOVE On the night shift. No 43140 is seen on 14 November 2004 waiting to be taken around to St Philip's Marsh depot after completing its work for that day. This was the final power car with white cabsides in the new First Group-style livery scheme.
(Chris Martin)

to involve notching them up more slowly and wait for them to react. Opening the power handle too quickly simply made matters worse and slowed acceleration yet further. If at all possible, the VP185 should be kept running above idle but with the typical stop-start nature of our services this was impractical. The only time one of those slugs were ever any use was when leading a train over really bad railhead conditions as they were far less prone to slipping. The weight of the rest of the train passing over the wet railhead cleaned it up sufficiently before the power-providing Valenta power car at the rear needed to use it!

The workload of a driver on the Great Western line was increased by the Automatic Train Protection (ATP) system in operation on that route. ATP had good points – the safety benefits of a system that prevented trains from approaching red signals or reductions in line speed too fast cannot be denied, but the system was flawed as it assumed the worst-case brakes on the worst-case rail conditions. This meant that when reducing train speed on the approach to a permanent or temporary speed restriction the brake needed to be applied much sooner than necessary in order to prevent a warning, or, even worse, an intervention by the ATP system. The ATP system was already using old technology when it was brought in. It relied on 'leaky feed' cables and track-mounted equipment for updates; the frustration of having to bring a train to a near standstill when you can see a signal in the distance which *was* red, but is now green cannot be overstated.

Changes to the structure of the rotas meant that within a couple of years I'd learnt the longer and more complicated routes including the line from Oxford to Hereford and from Reading to Exeter via Westbury, which had hitherto been the preserve of only the most senior/experienced drivers. Driving the IC125 away from the 'either full power or braking' environment of the main lines provided a far more interesting and rewarding driving experience. Neither route called for full speed running; instead (if running on time) the focus was on using the gradients to do much of the work for you. If power was applied or removed at the right points the undulating nature of the line meant that the train would coast down to speed reductions or naturally accelerate owing to a falling gradient. The nature of many of the more rural stops particularly on the Hereford route resulted in the need to remain vigilant to seek out more obscure braking points and the use of various seemingly antiquated mechanical signals and single-line tokens on that route added to the interest on that line.

Whilst some of my colleagues were less fortunate, during my time driving the IC125 I encountered very few failures; none that required assistance from another train anyhow. The first occasion when it looked doubtful that the destination would be reached was on 19 July 2003 when I took over the 10.15 Paddington to Weston-super-Mare at Bristol Temple Meads with double VP185s 43191/179. When trying to accelerate away from Bristol I found that 43191 was supplying ETS (Electric Train Supply) only, with no traction power while 43179 had power notches 3, 4 and 5 all isolated! We limped to Weston not exceeding 40mph (on a 100mph stretch of line) and the control centre very wisely decided to swap over onto a different set at Bristol on the way back, with 43023/152 substituting for the remainder of the journey to London. The only 'nearly failure' was on 8 October 2003; the set for me to work back to Bristol on the 19.15 from Paddington was late coming in from Swansea and station-based technical staff were attending (drivers generally describe all such staff by using the inaccurate catch-all term of 'fitters'). It eventually reached London with 43017 leading providing ETS only, 43185 would lead the train back to Bristol on the 19.15 departure. A quick examination revealed that the train would have to go back to Bristol with just 43185 providing traction power. The passengers were loaded as quickly as possible and off we went, but something was badly amiss. The usual notch 2 maximum from Paddington didn't matter as the rear one (under the station roof) wasn't providing power, so notch 3 was selected and a very slow start away began, applying more power into notch 4 slowed it down even more and the train came to a total stand still within 20 yards with 43185 in notch 5. A second attempt provided the same result, now with about two carriages off the platform at Paddington. After some discussion with the signaller and the intervention of a supervisor at Paddington it was soon decided to change cabs, bring the train back into the station and call the fitters back for a second look. One stood doing nothing except ranting and raving about how drivers don't know anything and will do what they can to avoid having to do any work; the other went in and quickly diagnosed the fault. No 43017 had failed to 'reverse' so the power cars were actually fighting against each other. Rather than risk a repeat failure a replacement train was found for our journey home to Bristol.

On 5 November 2005 (a time of year when track conditions are notoriously bad) I was driving 43030/024 on a Bristol to Paddington train which was diverted to run non-stop via Bristol Parkway owing to track repairs on the main route. The train had seemed poorly and acceleration was significantly worse than expected, but it felt similar to the sensation of when driving with poor railhead conditions, and nothing else untoward was noticed. Around 25 miles into the journey I was called up by the guard who was in a state of some anxiety. He said he would have to apply the emergency brake to stop the train as it was filling up with smoke. I told him he needn't as I would stop immediately. I made a heavy but controlled stop beside a lineside signal at Little Somerford where I could easily contact the signaller. As soon as the train stopped I opened the cab door and looked back. I couldn't see the train at all owing to clouds of smoke so I immediately shut down the engine

and made an emergency call to the signaller arranging for the opposite line to be blocked in case passengers started evacuating the train. After the smoke cleared sufficiently I could see that it was all coming from the bogies of the leading passenger coach. The brakes were stuck solidly on and probably had been since Bristol. It was then simply a case of isolating the brakes on that coach, the guard came and helped with a rotation test (a visual check to see the wheels were turning freely) and we carried on, albeit with that carriage locked out of use to passengers while the fumes from the brake pads cleared.

Another memorable day (but one which is best forgotten) was on 24 June 2006 when I had problems with different sets both ways. While driving 43181/136 up from Bristol to London on the 15.30 from Temple Meads I was stopped by the signaller at Milton, just west of Didcot. The driver of a train going the other way had reported clouds of smoke coming from my rear power car. No need for me to shut it down as it had died while I was coming to a stop and when I got to the back I found out why. No 43136 had a serious coolant loss and what the other driver had seen was the steam vapour from the escaping, boiled-over coolant. A different set coming back was managing without much help from leading power car 43012 as it persistently shut down upon restarting from station stops.

Complete failures were rare as the ability to limp on at reduced speed with only one power car working is a feature of the IC125, which is, maybe, relied upon slightly too often!

ABOVE No 43136 accelerates a West of England to Paddington train back up to line speed following the Exeter call during July 2006. By then, all power cars were wearing this simpler blue with pink relief and gold band version of First Group livery. (Steve Vaughan)

Chapter Five

On the road with a Technical Riding Inspector

Limited time for attending to the hard-worked trains during overnight servicing and the need to keep sometimes temperamental machines working properly, provided new challenges. Part of the solution was to create a small team of technicians tasked with riding on the trains to monitor and repair faults and to diagnose bigger problems for depot attention.

OPPOSITE No 43185 *Great Western* has just crossed Dockyard Viaduct as it nears Keyham, on the outskirts of Plymouth, whilst leading the 14.00 Penzance to London Paddington on 1 August 2018. This power car was reliveried back to the popular "swallow" livery as part of the 40th anniversary events in 2016. *(Chris Hopkins)*

The role of Technical Riding Inspector (TRI) was created early in the lives of the IC125 fleet. Rather than just waiting for the trains to fail and attend to them back at the depot, a small team of experienced engineers armed with a portable tool kit would ride on the trains themselves. This would allow them to check for problems which would often manifest themselves at speed when the trains were working hard, but be difficult to see and trace in a depot environment. They could also check the success, or otherwise, of attempted repairs to such faults that the depots had undertaken during the short time available to attend to them overnight.

Initially, the TRIs were generally static and based at key locations on the main routes, but upon the advent of mobile phones they could work far more flexibly as they could now be quickly contacted by the route control centre and directed to any errant train. Occasionally they could be diverted from looking at a minor problem and be dropped off by a train stopping especially alongside one in trouble so they could attend to the more serious problem and get things moving again as quickly as possible. This is the story of a few months in the career of life-long enthusiast and experienced maintenance engineer Tony Shaw as he helped look after a seemingly unloved pool of Virgin Trains Cross-Country IC125s in the months after IC125s had been replaced by brand-new trains on most routes.

It all started back in the summer of 2002 while I was working as a technician at Maintrain's Nottingham Eastcroft depot. I was asked if I would be interested in this role of a TRI as there was the possibility of a job coming up for somebody to ride around on the IC125s, looking after them on a day-to-day basis. Virgin Trains was beginning a new service called 'Challenger' and would be using six sets to run from Birmingham to Blackpool North and Manchester and would be needing three Technical Riding Inspectors (TRIs) to travel with them. Another couple of weeks went by before the Maintrain Fleet Engineer interviewed me, asked a couple of questions and got to know me a bit better. Then he said that Maintrain (the maintenance arm of Central Trains and Midland Mainline) would be looking at supplying three TRIs based at Birmingham to look after these IC125 sets for Virgin Trains. This position could last anything between three and twelve months and 'could I start next Monday?' I said 'yes' very enthusiastically! So it was straight home and sorted all my notes and circuit diagrams out from the attic so I could refresh myself. I had to pinch myself at the thought of being paid to travel and look after IC125s and listen to those 'Valenta' engines at full chat. Seven days to go: just in time for the new Autumn/Winter timetable which was starting next week.

Monday, 30th September, a day that will be remembered for a long time, certainly by me. Up bright and early and I made my way to Birmingham New Street for about 09.00. Fleet Engineer Mark was going to meet and introduce me to the other TRIs. After shaking hands with Phil from Birmingham and Mick from Derby, we collected our tool kits and were all issued with a copy of the IC125 'bible', the all-essential fault-finding guide. Never go out without a copy of it! The first day was to find our way about, meet the guys up in the control office, find where the station fitters hang out and, most importantly, work out what the trains were working that we would be looking after.

I wasn't allowed to ride in the cabs the first couple of days because the official 'cab passes' hadn't arrived yet. So I made full use of the time and went to meet the trains as they arrived at Birmingham and introduced myself to the drivers. I told them all that hopefully by

the end of the week one or all of us would be out travelling with them, but keen to get going, I did manage to blag my way to Birmingham International and Wolverhampton on the first day. Me being the computer nerd, I drew up timetables, diagrams, and set formations for the other guys.

I think it was Wednesday, 2nd October when I first rode out officially armed with a cab pass, tools and the 'bible'. I booked on at 08.00 with control and was asked if I could ride XC38 for a post exam audit. It was working 1P32, the 09.46 Birmingham International to Blackpool North. So I headed over to platform seven at Birmingham New Street to wait for it to come in; 43180/087 to Blackpool and back. XC38 was formed in a 2+5 set with 43180 leading, 44042, 42117, 42336, 40414, 41108 with 43087 bringing up the rear. Up to the front and into 43180 to meet the driver and inform him that I was travelling with the set all the way up to Blackpool. Later, I found out that most of the drivers welcomed a TRI on board as it gave them a little more confidence of getting there.

I didn't find too much wrong with the set after the exam, but the one thing I did find out very quickly is that a five-car IC125 set doesn't half move, especially when you are sitting in the back cab watching where you have just been. The 'test track' to see if you had got a good set was between Wolverhampton and

Stafford; you would expect to be doing 125mph by Four Ashes. Upon arrival at Blackpool I took my first chance to get a photo of one of 'our' Challenger IC125s at the resort. I was pleased to be paid to come here because Blackpool is one place I do enjoy with the vintage trams running along the prom.

Thursday, 3rd October and my first task was another post-exam audit, but on a different set. XC33 formed with 43159, 44060, 42187, 42326, 40432, 41095 and 43086 and a different train, 1H26 08.51 Birmingham New Street to Manchester Piccadilly. I could see myself enjoying this job! The following week started with a bang. On Wednesday, 8th October I was asked if I could ride XC32 with 43155/063 on 1P31, the 08.03 Birmingham International to Blackpool North. I only just got to Birmingham in time to catch it, jumping on at the last minute. It felt very sluggish so I donned my ear defenders and walked up to the front with 43155. I spoke to the driver; she said it felt like the back power car was dead so I made my way through the train to find out. When I got there 43063 was silent and cold, a quick look around it didn't reveal anything amiss. So I de-isolated the engine and restarted it at Stafford. All appeared OK when idling but when we departed it 'blacked out' Stafford station with thick exhaust smoke. I quickly turned the control cut-out switch to keep the engine idling.

ABOVE The plume of exhaust from No 43067 shows this power car still has the original Paxman Valenta engine fitted, as it begins the journey from London King's Cross to Sunderland with the 16.50 departure. Grand Central operated a small fleet of six power cars and three sets from 2007 until 2017 on this route.

(Alex Wood)

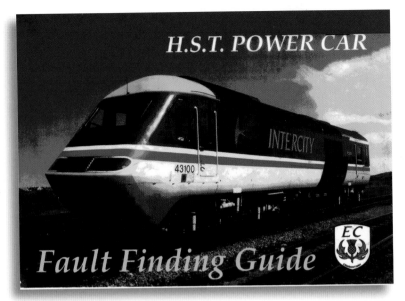

H.S.T. POWER CAR

INTERCITY

43100

Fault Finding Guide

EC

ABOVE The all-important fault-finding book: 120 pages to get you out of trouble with an errant IC125!

At Crewe, I asked the driver if she would depart on notch 2 until we had cleared the station then open her up, but the same again – this time we blacked out the Heritage Centre. I isolated the engine and rang Control who said: 'shut it down as the turbo was unserviceable, but can you now get off at Warrington and make your way back because 1G72 09.06 Blackpool North to Birmingham International was also on one power car'. No 43089 had shut down en-route so I jumped off XC32, across to the other platform to meet 1G72. Went straight to 43089 which was leading and found it was low on coolant so I rang the Crewe station fitters to see if they could meet me with a coolant bowser. Topped her up and restarted that wonderful Valenta and the train departed for Birmingham just 17 minutes late.

The next day, a different but less dramatic problem, as the driver couldn't get his driver's key out of the controller on 43194 – just the tool; WD40 works a treat. Next day I responded to a report of no saloon lights in TS carriage 42334 and met the train at Stafford on the 06.54 Blackpool North to Birmingham International. I managed to reset the BCCB (Battery Charge Circuit Breaker) at International and got the lighting back on for the return journey.

On 17th December 2002 a report came in that 1G52 was in trouble and could I go out to meet it. At first I was told it was a power problem, so bearing this in mind I managed to carry a set of power modules with me to Stoke-on-Trent. Waited for 1G52 to come in

only to find that 43193 was dead at the rear of the train. The driver said he had told control that it was shut down. So jumped on 43195 at the front and walked through the train to the back only to find that 43193 was low on coolant and had shut down. I could get coolant at Birmingham New Street but the engine was cold. I had one chance but only if I could get 43193 started and warmed up and we could get the coolant into it quickly during the now-reduced turn-around time in Birmingham. A technique is to wedge WLR (Water Level Relay) in, making the electronics think there is enough coolant for the engine to run, but this is normally frowned upon. But at Wolverhampton I went for it and got 43193 started and made the call to the fitters at Birmingham New Street to meet me with a coolant bowser. As I walked back through the train to the front to the driver an announcement was made by the Train Manager 'as you may have just seen and heard, our rear power car has been re-started by the fitter walking through the train'.

I started printing Set Formation sheets to show which of the 22 power cars were in service and which were not. The 'Challenger' fleet was allocated power cars 43063/069-071/078/079/086-089/155/156/159/161/162/180/193-198. Out of these we needed twelve power cars for the six sets to run a full diagram; three on the route between Birmingham New Street and Manchester Piccadilly, and three on the route from Birmingham International to Blackpool North. This is where having three TRIs helped because I always offered to do the Blackpool turns whereas Phil preferred the Birmingham area, and Mick didn't mind Manchester. That allowed me to visit Blackpool no less than eleven times in the three months up to the end of the year.

The 'Challenger' sub-fleet had been created by Virgin Trains when widespread use of their IC125s ceased on Cross-Country duties following introduction of their new four and five-carriage 'Voyager' trains during 2002. The idea was that the trains would have two standard class carriages removed and would then be formed with two power cars and five carriages. This fleet would be given full overhauls and the carriages refurbished to match the new trains and they would continue to work with Virgin Trains, but on new routes. They were intended to operate trains from London Paddington to Birmingham via Swindon with many of those services continuing through to either Blackpool or Manchester. Blackpool was a new route for Cross-Country so use of them on the sections north of Birmingham was a sort of preview of the new services. This meant the power cars chosen for 'Challenger' were those which had run the longest since their last overhauls as they would soon be receiving major attention, and often this showed. Although we only needed 55% of the fleet in traffic to run the service the neglected condition of some of the fleet meant even this low target took some work to achieve!

By late 2002, the plans to overhaul the fleet were finally dropped so in January 2003 the diagrams would be changing and we would be losing one Blackpool and one Manchester duty, but would gain the 06.44 York to Penzance (1V31), 08.30 Preston to Plymouth (1V35), 07.50 Glasgow Central to Bournemouth (1O16), and the 10.38 Bournemouth to Glasgow Central services (1S68).

So, onto 2003 which started as 2002 left off. On January 1st I was heading back to Blackpool with XC33 on 1P32. This time it was 43193/086 on the set and another post-exam audit. January 2nd, a different challenge as Control rang me to say 43194 was low on power with a fault light lit in the cab and could I meet it in Birmingham and have a look? Over to platform five and met it, 43194 was leading. After speaking to the driver he said he could not get more than 1,000amps before it cut-out. Straight to the cubicle, PEFR (Power Earth Fault Relay) was tripped, usually signifying a traction motor problem. I reset PEFR, but it tripped straight away when we departed Birmingham New Street on the empty run, which cleared the platform at the busy station by running a loop around Aston. All I could do was advise the driver to keep the power down to a minimum unless he needed to increase it, and then informed Control that 43194 would require attention at the end of the day. I didn't stay away from Blackpool for long during 2003 as I visited this fine seaside town a further 23 times up to the May timetable change, and the end of 'Challenger'.

So back to early 2003, 6th January to be precise, and I was diagrammed to go up to Blackpool with XC32 to do an audit on the set. Nos 43087/071 in charge and an excellent run was enjoyed up to Blackpool, arriving early everywhere requiring us to wait for time. On arrival at Blackpool my mobile went, Control wanted to know how much fuel each power car had. Why? Because they didn't know if they had been fuelled last night at Wolverhampton's Oxley depot! I was surprised to see both power cars only had just over one-quarter of a tank of diesel. Control wanted to cancel the return working but with a very enthusiastic driver we devised a plan allowing us to take the set onto Blackpool North fuel point; almost certainly a first and last time for an IC125 set. Control agreed to it and we set wheels in motion, literally. Driving very gingerly across the junction onto the stabling point we picked up a mobile radio so we could talk to the ground staff and get the set in position. The driver stopped 43071 on the fuel point to get it filled up. Around 2,200 litres later – perhaps the most that fuel point has dispensed in one go – it was filled so I got out and guided the driver as far as he could move the set towards the stop blocks. Unfortunately, 43087 at the back was about 10 foot short of the fuel hose and didn't get fuelled. So we decided to run with 43087 providing electric train supply only, thus preserving fuel consumption.

BELOW No 43155's dials viewed in the back cab of a 'Challenger' set running at full speed.

Towards the end of January I'd managed to travel on all bar 43079 of the 'Challenger' fleet power cars to Blackpool North. With the diagrams changing again to include other long-haul Anglo-Scottish runs I could see me missing out with it, but by pure chance on 30th January I was working the afternoon shift and was asked to travel up to Stoke to meet XC32 coming back down from Manchester on 1G54. Quickly, I worked out that I could make it with a ten-minute connection at Stoke. I headed over to platform five at Birmingham New Street and jumped on 43156/162 so I could alight at Stoke onto 43071/087 on XC32. It had been reported as the Stills boiler in the buffet was not draining away.

Soon sorted that out and back to Birmingham and as we arrived at New Street I could see 1P35 sitting in platform seven waiting to depart, but with the driver on his mobile phone to Control. I ran across (well, as fast as I could with a bag and tool trolley!) and spoke to him and he said that 43079 was not taking power – no amps on the ammeter in the cab. I said that I would ride with him to try and rectify the problem. We departed Birmingham New Street about ten minutes late and I went and had a look around the electrical cubicle to see if anything was amiss. Quickly found that the PCR (Power Control Relay) was not energised. Until this relay is energised then no power to the traction motors is available.

There are certain safety systems in place to stop the PCR energising and I made sure that these were all in order, then with a well-aimed finger, pressed the PCR in by hand. A surge of power as 43079 burst into life and took power, about 1,600amps. I went back into the cab to see the driver. He was 'over the moon', probably because he was on the way home and could get there far quicker now! But when we departed Wolverhampton it was the same again, PCR had not energised. Close inspection of the relay found that it was sticking (it was a known problem with these relays, so the guides and trunnions would be replaced if this happened) and with a little help it would energise fully.

WD40 to the rescue again, but I told Control that I had better travel up to Blackpool with him – I couldn't turn down the chance to ride on 43079 to Blackpool, enabling me to have travelled on the whole fleet to the resort, even though it had a genuine recurring problem and I could fix it if the fault happened again. Thankfully, it didn't so I booked it for a relay change when it got back to the depot that evening.

On 29th April I booked on for the morning shift with control and they asked if I could do an audit on XC36. Found out that it was coming down from Blackpool North on the first train then back with the 09.46 from Birmingham International. On XC36 were 43063/156; I had completed my audit by Warrington so I went and sat with driver up the front on 43156. At Blackpool, I nipped into town but when I got back the driver said he hadn't got any working headlights on 43063. Both headlights are wired in series so I checked both headlamps to be OK. The only thing it could be was the wiring between them. I couldn't do anything about that now so rang control who advised us to proceed to Preston with an emergency headlight on the front and turn the set around at Lostock Hall. So we ran empty to Preston and then out to Lostock Hall Junction, changed ends there to Farringdon Junction and reversed again with 43063 leading back into Preston so 43156 could work South normally. The same problem would happen when we got to Birmingham, so we arranged to have the train routed into Birmingham via Aston allowing 43063 to lead for the short run down to Birmingham

International, and would then be at the back for the return working.

By now we had lost some of the 'Challenger' fleet for refurbishment. Nos 43159 and 43180 went to Brush Works at Loughborough and 43071 and 43087 were on their way to DML Devonport. This was to allow some power cars to be made ready for their next home: the short-term Manchester to London St Pancras service dubbed 'Project Rio'. This would see the former 'Challenger' fleet form the bulk of an expanded Midland Mainline fleet maintaining a through service via the scenic Hope Valley during upgrade works on the direct West Coast Main Line. As it turned out, I was to follow the 'Challenger' fleet and take up a role as a TRI working on the temporary services, extending my involvement with these particular IC125s for a further 18 months.

During my eight months working with the 'Challenger' project you gained a feeling for some of the power cars. Certain ones used to be a pain, such as 43063 as I have mentioned above on two separate occasions, but I can think of two more problems it had that I had to attend. On 13th December when it arrived at Birmingham

with A6 'Con Rod' sticking out of the crankcase door, and again on 23rd March when I picked the empty train up at Burton-on-Trent (perk of the job!) and travelled with it to Birmingham, but as we were passing through Duddeston station we had an unsolicited brake application. It cleared as soon as the brake pressure dropped, which was strange but it did it again about a mile further on. The driver got on the phone to New Street Power Box to request a 'straight run' into Birmingham which we got – green signals all the way. No further problems then until the return journey back from Manchester when the same happened again. The driver couldn't get the brake to release but mysteriously it cleared when I slammed the engine-room door. Later, I found out that a loose cable terminal in the bulkhead connection box caused all the problems. No 43063 even caused problems when I wasn't there. One of the roof-mounted air filter intake louvre doors came open as she came into Birmingham New Street, which caused massive disruption due to the 25kV overhead lines having to be turned off before somebody could retrieve it. On the other hand, power cars like 43156 and 43162 never gave me any problems.

BELOW No 43070 is seen at Blackpool North having just arrived on the 11.46 Birmingham International to Blackpool on 16 October 2002. No 43078 *Golowan Festival Penzance* had headed the train into Blackpool.

(Chris Martin)

Chapter Six

21st century trains: how the IC125 was modernised

Running the iconic Inter-City 125 well beyond its expected service life resulted in the replacement of major components with newer equipment. This chapter looks at the development and introduction of new engines and cooler groups, along with other refinements which have allowed the trains to continue to provide reliable service.

OPPOSITE The Royal Border Bridge's 28 arches have spanned the River Tweed between Berwick upon Tweed and Tweedmouth since 1850. Here, we see a full Virgin Trains East Coast livery set with No 43257 at the helm passing over the magnificent structure while working the 17.00 London King's Cross to Edinburgh on 18 July 2016. *(Chris Hopkins)*

ABOVE The impressive station building at York had resounded to the sound of the classic Paxman Valenta-powered IC125s for three decades, but it was soon to end. No 43096 *Stirling Castle* is seen leading the 09.45 London King's Cross to Aberdeen with No 43115 *Aberdeenshire* at the rear, on 12 May 2007. Just a few days later, No 43096 was called into Brush works, Loughborough to receive one of the new, much quieter MTU R41 engines.
(John Tattersall)

Whilst the IC125 may have been viewed as a short-term solution to the need for faster trains in the 1970s, it became apparent that they were highly successful trains capable of providing front-line service for many years to come.

IC125 POWER UNIT ORIENTATION

From the perspective of engineers used to the original power car layout (with Paxman Valenta engine) all subsequent engines are considered to be the 'wrong' way round. An ISO specification was established for engine bank labelling such that sitting on top of the engine looking towards the free end, 'A' bank is left and 'B' bank is right. This means that when installed in a power car an MB190, VP185 and R41 engine has the 'A' bank facing the 'B' side of the power car and vice-versa. In fact, the Valenta preceded any standard and was opposite to the prevailing norm, so someone new to IC125 engineering would have regarded the Valenta as being 'wrong'! This has resulted in some confusion; the 'A' side of the power car is on the left when sitting in the driver's seat, regardless of what engine type is installed.

Key to this success was the reliability and robustness of the Paxman Valenta engines and their associated equipment. With such a large and highly utilised fleet any meaningful saving in fuel economy or component overhaul cost would translate into substantial savings when multiplied across the fleet.

Advances in engine technology made the late 1960s' Paxman Valenta (with its origins in a 1950s' design) seem outdated and the possible advantages gained by replacing the engines could not be ignored. As the trains continued into the 21st century the age of the major components was beginning to take its toll. Whilst engines and cooler groups would continue to operate efficiently if well maintained and regularly overhauled, there was a finite limit to the life of parts which had already operated for many millions of miles in the punishing IC125 environment. Additionally, governments were responding to scientific and public concern resulting in stringent exhaust emissions regulations being introduced. Although none of these applied retrospectively, the need to embrace cleaner, greener engines would soon become key to the IC125 story.

A major turning point in the history of the technical development of the fleet came in 2005 when two of the main IC125 operators (First Great Western and GNER) were both awarded new franchises. Instead of replacing the trains

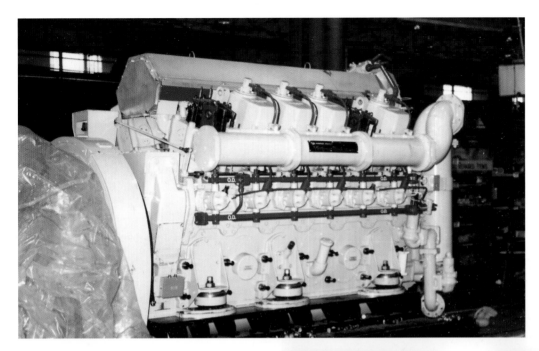

they were to be substantially overhauled with new engines, extending their service life for a further decade. Even when new trains finally began to enter service to replace the IC125s on the main lines from Paddington and King's Cross from 2017, around half the fleet was destined to continue operating on secondary routes, extending their lives yet further.

The Mirrlees Blackstone MB190 engine

A spate of failures on the Paxman Valenta engines led to BR engineers starting to seek a replacement power unit. Although it had generally performed well, the punishing environment of the IC125 work cycle had taken its toll on the Valenta and the benefits of installing a more modern and, in theory, more reliable engine were attractive.

Mirrlees Blackstone, the engine division of Hawker Siddeley, had just developed a new V12 high-speed diesel engine designated the MB190. The new engine differed from the Valenta as it had a 45° 'V' making the engine itself narrower but taller than the Valenta. It had a 70-litre displacement but was heavier than the Valenta. The MB190 was also more powerful with an output of 2,400hp vs the 2,250hp produced by the Valenta. This was seen as a significant feature which would allow the

RIGHT The Brown-Boveri turbocharger in situ at the top of the MB190 engine at the free end.

RIGHT A very rare view of the special Serck Behr cooler group made for the MB190 project. This will be on the 'A' side of the power car upon installation. The engine will be coupled to the cooler group to the left-hand side of this view.

BELOW The Heinzmann governor installed in a power car on the 'B' side wall next to the battery charger.

engine to spend less time running at full power. Operating the MB190 at 2,400hp reinstated an earlier abandoned aspiration to uprate the original Valenta engines to maintain schedules after an eighth carriage was added to some IC125 sets.

A Heinzmann electronic governor was installed which promised far more precise control than the hydraulic governor fitted to the Valenta. This itself would bring considerable benefits in reducing fuel consumption and exhaust output. The governor was mounted above the battery charger choke and had a rack frame with separate cards that dealt with loading and speed control. To avoid a costly re-design in the electrical cubicle a transducer voltage copying the output from the governor used on the Valenta was produced. Power supply to the governor originally came from a 24V convertor which proved troublesome, so an ingenious solution was found resulting in the installation of two large burglar alarm batteries above the cubicle. These provided an uninterrupted supply to the governor, and incidentally, this solution was adapted for another application when the Western Region fleet was fitted with a new electronically controlled safety system. Use of the electronic governor created a design problem for the engine over-speed rack protection and this was solved by using inlet manifold flaps which starve the engine of air following an over-speed incident.

A Brown-Boveri turbocharger was used and this was mounted at the free end of the engine resulting in all the exhausts being piped down to that end of the engine and returned through a large pipe to the original exhaust box. Although the turbochargers themselves proved to be very reliable, the amount of pipework around the engine was to prove a serious handicap for engineers working on it.

A significant problem with the MB190 was that it was incompatible with either existing type of cooler group so a new design produced by Serck Behr was produced, which forced up the conversion cost and brought in more new and unfamiliar equipment. The new cooler group comprised two fans and three cooling circuits and in a departure from the previous (and subsequent) cooler group types it was not symmetrical. The group was divided up vertically with a larger rear section consisting of two panels for the primary circuit while a smaller front section, only on the 'B' side, dealt with the secondary circuit. The 'A' side of the front section dealt with intake air. The intercooler function was dealt with by the cooler group itself using air cooling rather than water cooling and it was considered that relocating this item away from the hot engine improved reliability despite the extra pipework required.

In order to access the rear cooler group section a trap door was cut into the adjacent bulkhead wall. Each fan was controlled separately by an adjustable controller drawing data from probes. The controller was fed from the 24V supply used for the governor. As fan control was far more precise, cooler group efficiency was considerably better and it was found to be able to cope with higher ambient temperatures and dirty radiator panels, without encountering the issues which affected the earlier types.

The combined weight of the new installations made the power car 3.8t heavier than with the Valenta, adding nearly 1t to the axle loading of the power car. This was manageable with just a small quantity of power cars, but potentially a serious problem for the civil engineers if the whole fleet was to be converted. To mitigate the extra weight the bogies were shimmed up using extra spring packing on the primary suspension.

In December 1985, BR placed an order for five MB190 engines, four for installation in Neville Hill-based Cross-Country power cars (which BR was funding) and one as a spare (which Mirrlees supplied as part of the contract to support the trial). It was soon decided that whilst the Cross-Country fleet was exposed to the widest variations in operations including some 125mph running, it would be better to keep the new engines in a smaller operating area and working principally from just one depot. Western Region-based Nos 43167–43170 were selected for the conversion to be carried out at Derby Works. Before the power cars were called to works the first MB190 was lowered into No 43091 at Derby to test clearances and obtain data on the work required for the installation. The entirely different footprint of the MB190 necessitated extensive modifications to the engine bed plates. These modifications turned out to make it prohibitively expensive to reverse the process and reinstate Valenta engines. Nos 43167 and 43168 were converted during late 1986 and emerged for testing and in-service appraisal on 26 January 1987. The power cars entered traffic on 31 January and were initially confined to the Bristol to Paddington route with fitters on hand at Reading and Bristol to attend to any problems. No 43169 arrived at Derby Works for the conversion work in January 1987 with No 43170 following in February.

BELOW No 43301 leads a recently outshopped Cross-Country-liveried IC125 formation while working a Plymouth-bound service during March 2009. *(Jack Boskett)*

As the engine was taller and featured broad exhaust pipes along either side, access to the engine while it was in situ proved problematic. Attending to the power car engines while they were operating was made hazardous by the presence of the substantial hot pipes alongside the top of the engine. For depot-based maintenance work the restricted space available made attaching or detaching the cooler group to the engine drive flange particularly challenging. This required someone with long arms, high physical strength and the ability to work in a very confined space to work on this part of the engine. Removal, cleaning and replacement of a cooler group attached to a Valenta could be done in one shift, but only an especially adept fitter could just manage the removal of the cooler group attached to an MB190 during a single shift. Access to the bolts securing the big end was also notoriously tricky on the MB190. Anecdotal reports from the time they were maintained at Bristol Bath Road suggested they were highly unpopular to work on. Simply a rumour that one was coming in for attention would result in some staff going off work 'sick'!

Reliability was poor and one or more of the trial power cars could generally be found languishing in Bristol awaiting repairs. An early issue encountered came from cracks continually appearing in the corner mounting points of the cooler group. Repairs were made only for the problem to quickly reoccur and at one stage a decision was taken to try a further type of cooler group produced by Covrad. This was installed in No 43170 but it was found to be worse than the Serck Behr unit. It could not maintain correct coolant temperature as the primary circuit panels were too small, and the power car was limited to notch 4 as a result. In an attempt to cure the original problem, the designer from Serck Behr was sent to visit Bristol Bath Road depot to inspect the cooler group in person. Dumbfounded by what they found, they were on the point of giving up when by chance, the cooler group was viewed from above. It was then discovered that the fans had been cast the wrong way round. The cooler group had been unsuccessfully attempting to suck air from above the power car and expel it through the sides! Once this fault had been resolved reliability improved, but the engine itself had an Achilles heel: the valves had a tendency to break owing to extreme heat.

If operating at full power in notch 5 the exhausts would glow cherry red. Tests found them reaching 720°C and this would often cause a valve failure resulting in cylinder head and piston damage. To mitigate this the power cars were limited to only operate a maximum of notch 4 while alternative types of valve were tried. They still failed even when limited to notch

4, but less frequently. Power cars would often spend long periods out of service awaiting parts following such failures and this cemented the opinion that the engine suffered a lack of support from its suppliers. This was to ultimately seal the fate of the trials.

One particularly embarrassing occasion, when an MB190 failed, was on 14 June 1989. No 43169 had just received a level five exam at Bristol Bath Road and was freshly repainted in the current Inter-City livery. As the best looking power car in the fleet it was chosen to be named *The National Trust* in a high-profile naming ceremony at London Paddington. It made the journey up to London without a problem, but no more than a few minutes into the return journey it dropped a valve which ejected the contents of the oil sump on to the outside of the power car.

The higher power rating of 2,400hp made them popular amongst drivers and they proved to be very capable at putting in noticeably rapid performances, especially when paired together. Even with an MB190 at one end of the train a full 10mph could often be gained while climbing or accelerating. They were also considerably quieter than the Valenta, especially when operating at higher power outputs. In June 1987, a serious fire occurred on No 43169, which was found to have started at the base of the rectifier, but was not blamed on the higher

horsepower output of the MB190 by those close to the project. The damage was so severe that No 43169 had to be returned to Derby for repairs. There was a modification at the time to remove part of the suicide diode on the rectifier to prevent thermal events within the equipment. Consideration was given to down-rating the MB190, but this was found to be impractical due to slow loading if the top end of the rack was reduced, so they remained at 2,400hp.

With herculean efforts from a small team at Bristol, the four MB190 engines continued in traffic until 1991. No 43170 was the first to be side-lined in November 1991 following a major engine failure and No 43167 joined it in February 1992, dumped at the back of St Philip's Marsh depot. Prior to the engine failure on No 43167 it had been the best performer, but costly repairs could not be justified and the

BELOW No 43170 passes Hinksey Yard on 5 July 1991 with 1A15, the 06.05 Hereford to London Paddington. This elevated view shows the roof of the Serck Behr cooler group which was fitted to the four Mirrlees MB190-engined power cars. Four months after this photo was taken, No 43170 became the first of the four MB190 power cars to be withdrawn from traffic. *(Martin Loader)*

day after the first passenger MB190 run.

One interesting footnote came from the MB190 project. The success of the Brown-Boveri turbocharger prompted a further trial of these in five power cars. Nos 43126–43129 received new Paxman Valenta engines while No 43198 had a Brown-Boveri turbocharger fitted for one full overhaul cycle and their performance and reliability was found to be excellent. As part of the same trial, a Viking Series 22 governor was fitted to No 43128 and this gave far superior load control with less exhaust output, while the other three received updated governors similar to the original design.

Sadly at this time, the rail industry was going through major change so the funding to carry out a larger scale fleet modification was impossible. The four engines reverted to standard turbochargers at their first overhaul, although the three improved mechanical governors remained in use.

The Paxman VP185 engine

The next move towards a replacement engine for the Paxman Valenta arrived in 1993 when Paxman developed the VP185 engine as a successor to the Valenta. The new engine has a 90° 'V' with a reduced bore

ABOVE It was decided by the end of 1991 that any power car suffering a terminal failure to its MB190 engine would not be repaired. By mid-1992, both Nos 43167 (nearest the camera) and 43170 were dumped at the back of St Philip's Marsh depot in Bristol to await their fate. During this time, No 43167 was heavily robbed of parts to keep other power cars in service. (*Chris Martin*)

power car became a spares source looking increasingly unlikely to ever make a return to traffic. No 43169 remained in traffic with its MB190 power unit until March 1993. An attempt to repair the engine during an overhaul was doomed to failure. Examination found the sump to be overfilled with contaminated oil and removal of the manifold end caps revealed the remnants of the valves mixed with coolant. This left just No 43168 in service.

The life of the final MB190 engine was extended by isolating both notches 4 and 5 on the power controller until the final end came after completion of its duties on 31 January 1996, nine years to the

RIGHT A newly overhauled VP185 engine coupled to an alternator ready for installation into an East Midlands Trains power car in 2009. The embossed Paxman name and the VP185 wording cast into each rocker head cover have unusually been painted red, highlighting these items for an open day event at Neville Hill depot.

(185mm) and stroke (196mm) when compared with the Valenta, resulting in a more compact engine with a displacement of 63.2 litres. The 90°V made it significantly lower, but only slightly wider and longer than the Valenta, so would fit inside the power car after modification to the engine bed plate. The other significant change was to the turbocharger arrangement. Rather than one large turbocharger the VP185 was to use six identical, far smaller Schwitzer automotive-style units in a water-cooled exhaust collector box. There were two low-pressure turbos on each side and two higher pressure turbos mounted at the free end of the box. Each turbocharger was mounted in an easy-to-replace 'cassette' allowing them to be changed quickly.

The VP185 was capable of 2,700hp at 1,500rpm but was to be down-rated to 2,250hp in the IC125 to avoid undue stress on the rest of the equipment. Better fuel consumption and reliability was promised and the engine had clearly been developed with the rail traction market as one of its target customers. The VP185 uses a Viking Series 22 electronic governor supplied by Regulateurs Europa, enabling far more precise control over the engine. An associated control panel for this equipment is located in the luggage van area of the power car. At the time this was ground-

breaking as it was the first digital electronic governor fitted to any UK traction unit.

The electronic governor controls an actuator that operates the fuel rack on each bank of the engine. These racks operate combined pump injector units which are fired by cams on the single engine camshaft. Additionally, the new electronic governor incorporated improved engine protection by monitoring the primary coolant circuit and lubricating oil pressure/temperature. Engine power can be reduced if set parameters are exceeded, using an actuator on the fuel rack. A much later development has made an electronic version of the injector

ABOVE An unusual view of the VP185 in situ, taken from the space where the cooler group will be installed, and before the roof section above the engine has been refitted.

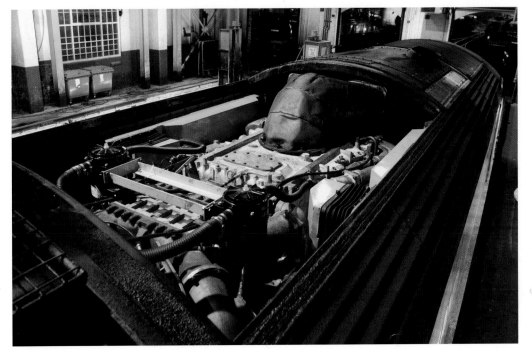

LEFT VP185 in situ, showing the turbocharger box, revised exhaust elbow, and air intakes. A later modification to around half the VP185s (including this one) introduced a closed-circuit oil breather modification whereby oil was captured by two black canisters and passed back into the engine.

137

RIGHT A close-up view of a VP185 with a rocker head cover removed, showing the valves.

FAR RIGHT The oil priming pump for a VP185 located on the floor on the 'A' bank side of the engine.

RIGHT A Bosch starter motor is located on the 'B' bank side of the engine.

available (see under the New Engine Governor heading below).

The first power car to receive the VP185 was No 43170 which had been side-lined for two years following the failure of its experimental MB190 engine. The work was funded by Paxman and carried out by Interfleet Technology, principally at Laira depot, during the summer of 1994. In order to gain accurate data from the new engine, an electronic diesel engine monitoring system (known by the acronym DEMON) which recorded 43 parameters on the engine was installed on No 43170. A similar but less complex system was fitted to standard Valenta-fitted power car No 43143 which had recently received an engine and cooler group

change. This was to allow comparative data to be gathered and as far as possible, the two would work together during the trial period.

No 43170 entered passenger service on 22 October 1994 on Great Western (GW) routes and initial fuel consumption and reliability data were encouraging, so it was decided to extend the trial by converting a second power car in 1995 for operation on the East Coast Main Line (ECML). Long-term stored MB190-engined power car No 43167 was sent to Crewe Works for conversion in January 1995, but the amount of work required to return the power car to operational condition postponed the ECML debut of the VP185 until 15 January 1996. While in storage at Bristol almost everything possible

BELOW A set of six VP185 turbochargers.

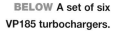

had been robbed from No 43167 to keep others in the fleet going. By the time it reached Crewe the only things left in the entire cab were a light switch and a defective windscreen wiper motor. The rest of the cab had been used as a template to make new cabs to effect collision repairs on other power cars.

Boosted by encouraging data from No 43170 and the desire to abandon the unsuccessful Mirrlees MB190 engine, Nos 43168 and 43169 also received VP185 engines, both visiting Crewe for this work, with No 43169 being completed in September 1995 and No 43168 in May 1996. Meanwhile, four power cars from the Midland Main Line (MML) fleet were to receive the new VP185 engine in place of their Valentas in late 1995/early 1996, and the conversion work was carried out at Neville Hill depot.

Serious problems beset the VP185s during late 1995 and 1996. These began with a catastrophic failure on the engine fitted to No 43170 at Laira in September 1995 when a build-up of carbon on the upper piston rings expanded and locked the pistons to the liners. This resulted in an engine seizure and the power car spending 13 weeks out of traffic while the engine was rebuilt.

ABOVE LEFT The actuator mounted on the engine actions the signals sent from the governor.

ABOVE The substantial Regulateurs Europa governor mounted on the bulkhead wall in the luggage van. It was fortunate that this space was free to locate this equipment away from the hostile environment of the engine itself.

When No 43047 (powered by MML's first VP185 engine) entered passenger traffic on 23 November 1995 it failed on its maiden run and it was found that an assembly error had resulted in a complete loss of exhaust pressure. The power unit was quickly repaired but a few weeks later a serious over-speed incident occurred when the power car was restarted after two days out of use, resulting in serious damage to the alternator. During testing following the alternator change, No 43047 suffered an overheated main bearing following the loss of oil feed due to unseen damage sustained during the overspeed event. This led to engine seizure requiring it to be returned to Paxman for rectification. Mechanical over-speed trips were fitted but these were prone to false trips so as a temporary measure it had been propped open by a piece of balsa wood, which was supposed to break during a genuine over-speed event.

Following that failure the VP185 was modified with an electronic speed switch that operates flap valves in the engine air intakes, thus starving the engine of air. No 43075 received the second VP185 engine for MML, but the installation of the third and fourth MML VP185 engines was postponed because of the ongoing problems. Reliability appeared to be recovering until 7 April 1996 when No 43169 sustained a spectacular fire following a rocker arm fracture which allowed unburnt diesel to fill the exhaust manifold and turbochargers.

During autumn 1996 the commitment to convert a further two MML power cars was fulfilled when Nos 43059 and 43074 received VP185 engines at Neville Hill. Despite all the problems that had been encountered, there was little choice but to pursue the replacement of at least some Valenta engines as the pool of the original engines was being gradually diminished by unrepairable failures. Four further GW power cars received VP185 engines in place of their Valentas in 1997. In order, Nos 43177, 43191, 43173 and 43179 received the new engines between April and August with conversion work being done at Laira. However, No 43173's career as a VP185 was cut short by a catastrophic accident in September 1997.

Now with an active fleet of eleven VP185 engines (six for GW, four for MML and No

43167 on its own on the ECML by then branded GNER) the fleet settled down. Whilst being an essentially sound engine, a lack of management focus gave the railway operators the impression there was insufficient technical support from its manufacturer. The railways' own engineers had amassed unrivalled experience in looking after the earlier Valenta, but it would generally take a lot longer to repair any problems arising with a VP185.

The engine had suffered excessive vibrations during early operation due to the design of the shaft between the alternator and the engine, which was 300mm long to make space for the charge air intercooler. The problem was partially mitigated by the addition of stiffening struts between the engine block and alternator flange, and later a stiffer drive coupling between the crankshaft and alternator. Early fuel consumption savings of between 12% and 15% were found to be somewhat optimistic as comparison was being made with a Valenta working on the other end of a train to the VP185. It was later understood the 'power curve' on the VP185 meant the engine took far longer to reach full speed when compared with a Valenta and so during acceleration of a train with a Valenta and VP185 combination, the Valenta was doing more than half the work and thus burning more fuel. A pair of VP185

RIGHT The later version of the Regulateurs
Europa governor was mounted in a revised
position in the rear corner of the luggage van, in
power car '48'.

engines on the same train would generally
result in much slower acceleration of the train
as the governors were set to match the original
specification of the Valenta, which had normally
been beaten in the power car application. Initial
performance from No 43167 proved to be an
exception to this until it was discovered (too late
to prevent damage elsewhere on the power car)
that the power control system had gone awry.

Work was later done at Brush,
Loughborough during installation of new traction
control electronics to alter the governor settings
to make the VP185 more responsive. Following
trials using No 43075 a refined setting was
developed and applied across the fleet resulting
in an engine better suited to the stop-start
nature of the rail application.

Changes to the ownership of Paxman and
the realisation that no substantial orders for
the VP185 would be forthcoming from the rail
industry, only served to deepen the lack of
technical support the engine was receiving. In the
summer of 2001, GNER converted No 43167
back to a Valenta engine during an overhaul at
Doncaster Works in order to standardise its fleet.
The surplus engine was installed in No 43165 at
Laira in late 2001 for the company now branded
First Great Western (FGW).

Substantial maintenance cost savings were
possible with the VP185 when compared with
the Valenta owing to it being able to operate for
double the time between overhauls. The VP185
required a half-life overhaul every three years
(12,000 hours) and a full-life overhaul every six
years (24,000 hours). This motivated Midland
Mainline to work with Porterbrook (owners of the
bulk of their power car fleet) and convert a further
14 power cars from Valenta to VP185 between
2002 and 2004. When added to its existing
four this would result in 50% of the projected
36-strong fleet running with the VP185.

The conversion work was carried out by
DML, Devonport and power cars Nos 43043-
045/048-050/052/055/060/061/072/073/076/
082 received VP185 engines. While FGW was

LEFT Marking its role
as a VP185 test-bed,
No 43170 was named
Edward Paxman after
the son of Paxman's
founder, who created
high-speed diesel
engine predecessors to
the Valenta and VP185.

BELOW A self-evident fault: No 43045 displays what happens when a
turbocharger fails. Being one of the highest stressed components in a large
diesel engine, failures are not that uncommon in all engine types, but with
the VP185 changing them is quick and easy. *(Paul Corrie)*

trialling the alternative MTU engine, two VP185 engines (those installed in Nos 43170 and 43179) were modified, receiving new injectors, updated control equipment and single-bank firing while on idle. Neither power car received any attention to the long-standing issue regarding the governor settings so it came as little surprise when they decided to dispense with the VP185 engines in their fleet and convert all power cars to MTU engines.

When the seven FGW VP185 power cars were converted to MTU engines as part of the fleet-wide "repower" programme (see below) the opportunity was taken to convert the remaining power cars in the now slimmed-down fleet for the Midland Main Line route (now branded East Midlands Trains), so during 2008/09 power

cars Nos 43046/054/058/064/066/081/083 received VP185 engines at Neville Hill. The last power car to receive a VP185 engine was East Midlands Trains' No 43089 which was converted by Neville Hill in 2010, completing the fleet. Earlier conversions of Nos 43072 and 43074 left the fleet in 2011 and were given MTU engines by their new operator in order to standardise its fleet. This left 24 power cars with VP185 engines which can generally be found working on the routes from South Yorkshire and Nottingham into London St Pancras.

The Voith cooler group

During 2000, two First Great Western (FGW) power cars Nos 43023 and 43132 received experimental new cooler groups manufactured by Voith. Although there were no widespread reliability issues with the existing cooler groups, it was determined that the Serck cooler groups were becoming life-expired and their inherent design flaw allowing engines to continue operating after the fan had failed was becoming intolerable. Additionally, neither type of cooler group could cope with higher ambient

LEFT The diversion of trains away from the Great Western Main Line resulted in the unusual sight of an IC125 approaching Brookwood (near Woking) on 31 March 2013. The train is 1O36, the 07.53 Exeter St Davids to Waterloo with No 43027 *Glorious Devon* leading and No 43021 *David Austin – Cartoonist* at the rear. *(John Tattersall)*

temperatures which resulted in frequent High Water Temperature (HWT) relay trips during summer months causing delays to trains.

The two experimental Voith cooler groups were built using parts from the original Marston Excelsior design, but featuring two large vertically mounted radiator panels with each side comprising two panels. These were the outer panels for the secondary cooling circuit and the inner panels for the primary circuit, as per the Serck design. The plumbing arrangement was redesigned to use fewer hoses and connections in order to reduce possible sources of coolant leaks. The roof of the cooler group was identical to the original Marston design with lightweight slats being opened by air-flow from the fan and closing by gravity when no cooling is required.

A more powerful and robust pulse-width modulation solenoid operated hydrostatic fan with improved oil seals was installed at the top of the cooler group and this was controlled using an electronic control panel allowing much better control over the fan operation, which was intended to overcome the flaw with the Serck design. The electronic control panel was located in the adjacent engine compartment and this monitored three temperature parameters: the primary circuit temperature, the secondary circuit temperature, and the ambient air temperature. By monitoring the actual outside temperature it could react by compensating for changes, enabling the fan speed to be increased during hot weather before the coolant temperature reached the range at which cooling

ABOVE The second version of the roof panels fitted to the Voith cooler group with half-length flaps hinged in the centre.

BELOW An unusual view of a Voith cooler group in situ while the exterior radiator grilles are removed for corrosion repairs.

would be required. Conversely, in cold weather or soon after first starting the engine, the cooler group would delay operating, enabling the engine to reach the optimum temperature.

The control panel was set to only react to the primary circuit temperature while the power controller on that power car was at idle. This limited the previous problems caused when cooler groups would continue running at high speed when not actually required, as the secondary circuit was triggering unnecessary cooling demand. Electronic control allowed for a more sophisticated safety protection system. If an initial over-temperature parameter was reached a relay would prevent full power (notch 5) being taken on the affected engine and if a second over-temperature parameter was reached the engine would revert to idle until the system recovered, as it would on the earlier cooler groups. If the control system failed (such as in the event of a power loss, sensor fault or processor fault) the cooler group would operate at maximum fan speed until the fault was cleared or the system reset. Fan speed was controlled by means of electronic signals to a pneumatic actuator unit which varies the air pressure to the fan coupling between zero (fan off/idle) and 5.5bar (fan operating at maximum speed).

Following trial operation, 35 of the new Voith cooler groups were ordered in late 2000, but problems with the control systems on the two experimental power cars (solved by relocating the control panel to the luggage compartment) delayed their introduction until Summer 2002. Initially, numerous problems were encountered with the production examples supplied by Voith. Control system defects arose which resulted in most of the cooler groups operating at full speed, and then there were a few occasions when the cooler group shut down without shutting down the engine, resulting in serious damage. A problem with the design of the roof-mounted vents led to several power cars running with the roof panels missing until a revised design was produced.

The reliability issues were tackled by Voith and soon the new cooler groups were operating well. Installation continued to encompass 34 FGW power cars by the end of 2003 leaving one spare Voith cooler group and reducing the number of Serck cooler groups in use to just 12 examples. Some further problems were encountered with the Voith equipment in 2005 requiring a further re-design of the roof slats. These were replaced with a fixed roof grille arrangement protecting the fan from debris entering from above. Keeping the new Voith cooler groups clean proved problematic as the radiator fins were much narrower than the previous types and the gap between the primary and secondary panels would collect dirt. To combat this a new cleaning method had to be devised. When the new MTU 16V4000 R41 engine was installed into FGW power cars, the engine's own control system replaced the sometimes troublesome Voith control system, which further improved reliability. Additionally, the secondary coolant pump was no longer needed as, unlike the Valenta, the R41 engine incorporated its own engine-mounted pump.

The Brush cooler group

Astudy had been carried out on the Midland Mainline fleet and one of the key weaknesses identified on its power cars was with the cooler group. Even when clean, the original units were barely able to cope with high ambient temperatures. The Brush cooler group was initially conceived as a means of producing a radiator of more than adequate performance and to extend the period between overhaul without the considerable cost of building entirely new cooler groups. No 43051 was modified in 2000 using modified Marston-type radiator panels and a new fan system, but it was realised by Brush after testing, that an entirely new design using experience gained producing new cooler groups for Class 57 locomotives, would be needed for the IC125 power car.

The radiator panel layout used on the new Brush cooler groups is identical to that used on the Serck and Voith cooler groups, but the panels are substantially larger to increase cooling capacity, and are combined units without gaps between the primary and secondary panels. In a significant departure from previous designs the Brush equipment uses a pair of large roof-mounted fans which maximise air flow around the radiator panels. The hydrostatic fans are driven using a pump connected to the free end of the engine by means of a cardan shaft, and the fan speed can

ABOVE A Brush cooler group awaiting installation

BELOW A Brush cooler group with radiator panels removed, exposing the innards.

RIGHT Viewed from inside showing the two-fan configuration which gives the Brush-design cooler group superior cooling capacity when compared with earlier types.

be fully modulated to allow maximum efficiency (although in the Midland Mainline application with 2G or Viking governors conventional wax stat controllers are used). As with the Voith cooler group, electronic fan speed control is used. An electrically driven secondary coolant pump (supplied by Weir) is used when the cooler group is coupled to a Valenta or VP185 engine. This is not required on the R41 as it has its own engine-mounted coolant pump.

The secondary pump is powered by a feed from the three-phase and is able to operate whenever the power car engine is running. The cooler group is larger than the previous types making for a very snug installation with no void space between the cooler group and bulkhead. The extra size and an all-steel construction result in significant extra overall weight. The Brush cooler group fully uses all the available space and includes a larger header tank. As a result, the maximum permitted weight capacity in the adjacent luggage compartment was reduced by 1.4t, but this posed no operational problems as the luggage area was never used to full capacity.

Additional weights have been welded into the luggage van pockets to perfectly rebalance the power car such that each primary suspension spring bears an identical load. Four Midland Mainline power cars received the new Brush cooler groups during 2004 to trial the system. Following the success of those trials a further eight MML power cars received the new cooler groups by the end of 2005, along with all three power cars in the Network Rail fleet. Two power cars from the GNER fleet (Nos 43099 and 43102) received the Brush cooler groups in late 2005, which became a precursor for fleet-wide installation of this equipment on the whole ECML fleet. This was despite investment in new radiator panels for all the Marston-type cooler groups in that fleet only a few years previously.

The rest of the MML fleet was to receive Brush cooler groups, which ousted the previous types from the entire IC125 power car fleet nationwide. The Brush items became the choice for all operators, except First Great Western during the 'repower' programme detailed below.

Cab reliability and comfort upgrades

Although an apparently largely cosmetic change, the need to renew the driving cab desks was driven by three major factors: the amount of equipment added retrospectively, the obsolescence of the older pushbuttons, and the desire to improve the reliability of the ageing controls. Power cars for the ECML fleet received upgraded cab desks during a refurbishment between 2001 and 2003, although changes were largely cosmetic, with modern plastics and paints providing a better environment for the driver, but with most switches/controls staying unchanged.

In 2005, Brush, Loughborough developed a new cab desk which was designed to be added to the original framework. The previous cab desks had received little attention since the fleet was constructed and age-related failures of some of the controls were becoming an increasing concern. LED lights replaced the old, unreliable filament lamps, gauges for the brakes, speed and air system were angled towards the driver, and all gauges were now provided with dimmable backlighting. All retrofitted items such as the Drivers Reminder Appliance (DRA) were incorporated within the main bank of switches and the Automatic Warning System (AWS) sunflower indicator was relocated to a smaller unit directly in front of the driver, taking up part of the old clipboard area.

Indication lights for the Train Protection and Warning System (TPWS) were relocated within

ABOVE A Brush cooler group from above.

BELOW A GNER modified cab desk with improved dial angle and illumination, as shown to good effect here. The top of the dashboard is becoming cluttered with retrofitted items. Below the right-hand corner of the windscreen is a crew-to-cab call system fitted only to GNER power cars.

RIGHT **Work in progress to install the new cab desk in No 43480 showing how it was grafted on to the original framework.**

ABOVE The completed installation on No 43467. On earlier power cars the switch panel at the bottom left of the photo was largely unchanged, but the need to install the new GSM-R cab-shore radio system resulted in that panel being revised and raised to incorporate the new radio. Forward-facing CCTV has been installed at the bottom right-hand corner of the windscreen.

LEFT The left-hand side of the cab in No 43159 shows the position of the GSM-R handset to the right of the soon-to-be-redundant NRN cab-shore radio unit. Below these is a mimic panel for the frontal lighting.

the drivers' eye-line as mandated by new industry regulations. New isolation switches for the DRA and the new electronic vigilance device were relocated out of reach on the far right-hand side of the desk. Lighting control now includes a hazard-warning pushbutton which allows the headlights to flash alternately to give warning to other trains during an emergency. Each panel of the new desk layout was fitted with a 'Litton'-type plug and socket arrangement replacing the previous hard-wired arrangement, making repairs far simpler and quicker. The first power car to receive the new Brush-style desk was No 43013 and the fitting was later applied to all power cars, albeit with further modifications on the FGW power cars to incorporate equipment associated with the Automatic Train Protection (ATP) system in operation on its lines.

A general programme of changing the driver's seat with an improved, more adjustable design

had been carried out progressively on a fleet-by-fleet basis. One of the other significant driver comfort issues was addressed by replacing the cab doors to reduce the amount of wind noise encountered at speed. Two different designs were introduced. One was a door with no opening window and a small flap located at the base of the door for ventilation, which was fitted to power cars operating for First Great Western, East Midlands Trains and on the New Measurement Train. A second design, with an opening window hinged in the centre, was fitted to the rest of the fleets. The window would normally be locked shut by means of a 'T' key, but could be opened if required.

ABOVE LEFT A slow-speed control function has been added on some power cars for carriage washing.

ABOVE A close-up of the buttons showing the re-labelled 'Local Engine Stop' which used to be 'Fire Alarm Test', but had always performed the function of shutting down the local engine. The 'Fire Alarm Test' function was relocated to a new button positioned on the face of the electrical cubicle.

FAR LEFT The two different cab doors are seen in this view of Great Western's No 43003 and Cross Country's No 43303 at Plymouth. (Chris Martin)

LEFT Inside of the cab door showing the openable window. This was in a Grand Central power car.

ABOVE Network Rail's New Measurement Train (NMT) is seen on the West Coast Main Line passing Tamworth on 9 August 2016 with No 43062 *John Armitt* leading but with No 43013 nearest the camera. This is an innovative use of redundant IC125 vehicles which have created a train capable of 125mph which is loaded with sophisticated track-monitoring equipment. It tours the network assessing the condition of the track in minute detail, spotting defects before they become serious problems. This train was running as 1Q28 from Derby to Litchfield, London Euston, Crewe and back to Derby. *(Lucas Walpole)*

Great Western ATP and DAS systems

BELOW Automatic Train Protection (ATP) speedometer and associated controls on a Great Western power car.

Installation of Automatic Train Protection (ATP) on the Great Western power cars has created significant technical differences. This equipment was fitted to a total of 123 of the 197 production power cars. A trial ATP system was developed in the early 1990s and implemented on the main lines from London Paddington to Bristol. This system was designed to prevent a train from passing a signal at danger and to prevent a train from operating in excess of the permitted line speed. The system uses track-mounted equipment to relay data to the train which interprets this to produce an in-cab signalling display, along with the ability to apply the train brake if required.

Additional equipment such as the Hasler speed probe, the ATP antenna and the ATP control system box, have already been seen in Chapter Two, but the two most visible changes in the cab have been the replacement of the standard speedometer with a revised version incorporating ATP indications and controls, and the installation of an ATP data entry panel. On the original installation the wheelslip and engine-stopped indicator lights (items 24 and 25 in the cab description in Chapter Two) were relocated to accommodate the enlarged speedometer which occupied the full depth of the dashboard. However, when the new cab desks were installed a much neater installation was provided.

Rather than being offset to the right, the speedometer and associated controls were relocated directly in front of the driver. The special ATP speedometer has a range of up to 140mph. A series of green LED lights around the full range of the speedo show the 'demand speed' (the current maximum speed or the new target speed if a speed reduction is required), and a series of yellow LED lights,

from 0-50mph, show the 'release speed' (the speed which can be maintained when approaching a previously red signal ahead of the system updating). The 'display window' below the dial shows dashes, dots and zeros in a seven-segment display dependent on signal conditions, and can also show three-letter messages in relation to signal/train operation.

Below the speedo are four buttons which control other ATP functions. From left to right these cancel a train-induced brake demand once the speed has reduced sufficiently, or the system has updated to permit this, allow a signal to be passed with authority from a standing start, allow the train to operate in 'shunt' mode for depot movements, and enable the system at start-up. To the right of the cab is the ATP data entry panel. This is where information about the train characteristics are entered before each journey. The driver must enter their ID number and then confirm or amend details of the train, such as the number of vehicles in the formation and confirm there are no running restrictions such as broken windows, isolated brakes, collapsed suspensions, etc. If there are any such restrictions then the data must be entered into the ATP equipment using the codes provided. This then gives the ATP system data for the

purposes of calculating braking distances, maximum train speeds and the length of the train when leaving permanent/temporary speed restrictions. This system came into full operation in Autumn 1999 after a period when the system had fallen into disuse. This was ended following serious collisions involving IC125s in September 1997 and October 1999.

Also on Great Western power cars, a Driver's Advisory System (DAS) has been installed between the ammeter and the ATP data-entry panel. This consists of a touch screen where the driver inputs the train headcode at the start of the journey. DAS locates the position of the train using a GPS antenna located behind the horn grille and compares the train position against the timetable. The screen display shows the estimated time and scheduled time at the next passing point or station stop allowing the train to

ABOVE Several First Great Western power cars received colourful advertising liveries from 2014 onwards. No 43192 is seen promoting 'Bristol 2015 European Green Capital' as it leads 1A06, the 06.20 Weston-super-Mare to London Paddington, passing Compton Beauchamp on 5 May 2016. (Martin Loader)

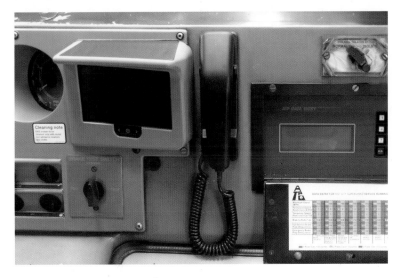

RIGHT The black screen located on the left is for the Drivers Advisory System (DAS) and to the right is the ATP data entry panel shown in the open position.

ABOVE First Great Western introduced a new image in 2015 using a dark green livery with traditional-style GWR lettering. No 43187 is seen freshly turned out in the new scheme having just arrived at London Paddington after working 1L74, the 14.54 from Swansea, with No 43188 at the rear, on 4 October 2015. *(John Tattersall)*

BELOW A brand-new MTU R41 engine prior to installation in a Great Western power car.

be driven more efficiently. To save the expense of removing the ATP and DAS systems they initially have been retained on the former Great Western power cars which are now part of the ScotRail fleet, although as there is no trackside equipment anywhere on the routes used, the ATP equipment now has no effect on train operation.

The MTU 16V4000 R41 engine

Smaller engines supplied by German manufacturer MTU had been used with varying degrees of success in some underfloor diesel multiple unit (DMU) applications for several years, but in 2004 it was announced that two IC125 power cars would receive the 16V4000 R41-type engine as an experiment. It was originally conceived that these engines would be used to assess the suitability and

reliability of the type ahead of design and construction of a replacement for the IC125, which was now approaching 30 years in front-line service, as well as evaluating MTU as a supplier of large engines.

The R41 engine is a 16-cylinder unit with a 90° V and a 65-litre displacement featuring smaller pistons, and a reduction in both bore and stroke to reduce stresses within the engine. Although the displacement is only slightly larger than the VP185, with its additional four cylinders, the R41 is somewhat longer than all previous IC125 engine types and installing it in the power cars required the alternator to sit further forward on an extender shaft to clear the intercooler. This all but eliminated the passageway around the rear of the electrical cubicle.

Movement of the alternator also required a modification to the alternator air outlet on the power car floor, and the false bulkhead between the engine and clean air compartments was moved forward. The alternators were slightly modified with an additional flex plate coupling on the crankshaft. Alternators modified to mate with the R41 engine were reclassified as BA1001C to distinguish them from unmodified units. The engine has a maximum rating of 2,600hp running at 1,800rpm and so was considerably down-rated to 2,250hp at 1,500rpm to match the output of the previous IC125 engines to avoid the need to make significant changes to the rest of the traction equipment. The engine uses a high-pressure common-rail direct-injection system offering far more precise injection control, thus reducing fuel consumption as well as exhaust emissions. The engine is equipped with four conventional turbochargers.

The engine is fully electronically controlled using MTU's own Diesel Engine Control system

(abbreviated to MDEC). The control panel for this is located in the luggage van on the cooler group bulkhead wall. Learning from one of the main problems encountered with the VP185, the MDEC is set to mirror the power curve of the original Valenta engine with each power 'notch' being reached in precise increments in the same amount of time the Valenta was originally set to achieve. This is apart from notch 4 which is increased from 1.350rpm to 1,420rpm.

The MDEC also controls operation of the cooler group and thus only an electronically controlled group can be used in association with the R41. Fortunately, the Voith cooler group (see above) had been successfully operated in service with First Great Western, so the two trial power cars were equipped with the newer cooler group and MDEC replaced the separate control unit for the cooler group. The original MDEC box incorporates three gauges showing engine speed, coolant temperature and lubricating oil pressure. The engine speed gauge shows speed from 0 to 2,000rpm in 100rpm increments. The coolant temperature gauge (labelled 0 to 120°C) and the oil pressure gauge (labelled 0 to 10bar) are both smaller than the engine speed gauge. Engine hours are logged using an analogue meter counter which is reset when the engine is changed. Three lights on the MDEC illuminate yellow if the engine encounters a non-fatal fault, and red if the engine has shut down due to a fault. On later installations (all except CrossCountry) the analogue gauges are replaced by an LCD screen upon which faults are displayed enabling much quicker and more accurate fault diagnosis. The LCD screen was retrofitted to earlier conversions to standardise the fleets.

One significant new feature with the R41 was that the engine would not start if the coolant temperature fell below 40°C. This resulted in the provision of an electric pre-heater on the 'A' bank side of the engine which uses power from the Electric Train Supply (ETS), either from the other power car or a static shore supply, in order to raise coolant temperature if necessary. The pre-heater is provided with its own circulating pump and elements along with a small local control panel showing the status of the system. A switch was added to the control cubicle allowing the pre-heat function to be over-ridden in the event that no ETS could be supplied to the power car, enabling the pre-heater to operate, but operation of the switch would only be authorised as a last resort.

Additional engine wear encountered if starting the engine from cold was such that 100 hours would be added to the electronic counter for that engine in the event of a cold start, meaning it would become due for overhaul sooner. New lights on the control cubicle indicate a pre-heater fault and if pre-heating is required, for which two additional switches have been installed. These are a pre-heat switch with 'normal' and 'over-ride' positions, and an emergency engine-stop switch, which can be padlocked into the 'stop' when the power car is being worked on. Additional circuit breakers were installed to protect both the 110V heater contactor and switch gear and the 415V heater supply. A

ABOVE The curtain separating the engine and clean air compartments in its new position following an R41 installation.

LEFT The MDEC panel located on the luggage van bulkhead on an R41-fitted power car. This is the first version with analogue gauges.

dipstick and low-volume oil filler are provided on the 'B' bank side and local start/stop buttons are located close to their original location on the Valenta. There is no oil priming pump so the R41 (provided coolant temperature has been maintained) can be started much more quickly than previous engine types. The R41 has a mechanical fuel pump driven by the engine to generate high-pressure fuel supply for the injectors. A later modification means this fuel pump is only used during engine starting, leaving the engine-mounted pump to draw fuel from the tank. Local air filtration is provided using two cassette-type filters on either side of the engine, eliminating the air filters located at the top of the power car. The bodyside air intakes are no longer used for direct engine air supply.

The two power cars chosen to receive the R41 engine experimentally were Nos 43004 and 43009. Both were previously stored as surplus to requirements although No 43004 had switched places with No 43005 as the latter was in far better condition. They were dispatched to Brush Traction, Loughborough and after completion, No 43009 was unveiled to the press on 11 May 2005.

Following test running, No 43009 was the first to enter service, on 14 June, and both power cars operated with FGW for evaluation. With scarcely enough time to accumulate sufficient data and experience of the new engine, both FGW and GNER (then operator of the East Coast Main Line services), announced commitments to replace the engines in their IC125 fleets as part of new franchises.

For FGW this was to include a number of previously stored vehicles enabling the fleet to expand to operate additional routes and oust unsatisfactory newer trains. Although early trials of the R41 had gone well, FGW took a considerable gamble by announcing that fleetwide installation of the still-unproven engine would commence in March 2006, with 26 power cars. These would be converted to R41 engines by Brush Traction and they would use Voith cooler groups already in service with the TOC. The scheme was to be termed 'repower' and in parallel with a carriage

'refresh' programme was to effectively provide a substantial life extension for its IC125s.

Within weeks it was announced that Brush Traction would also be converting 27 power cars to R41 engines for GNER, and these were to receive Brush cooler groups as part of the work. GNER had initially been expected to choose the VP185 to replace the engines in its IC125 fleet and had even considered the installation of entirely untried Cummins engines before selecting the R41. As the summer of 2006 progressed orders were announced by FGW for a further 87 Voith cooler groups. This provided enough cooler groups including spares for the whole fleet and paved the way for the conversion of the remaining 89 power cars to R41 engines. This made up a total of 115 power car conversions during the main programme for FGW, taking their fleet to 117 with the two trial conversions. The choice of R41 with Voith cooler group was unique to FGW, but this did at least standardise both the major components across its entire fleet for the first time ever.

The R41 uses a different specification of oil and coolant to previous engines. So, to ensure the correct consumables were added by the depots, the external fillers on all FGW power cars were temporarily fitted with adapters preventing accidental use of earlier-specification oil and coolant types during the transitional period while the fleet was being converted.

Brush Traction had previously only hosted power cars in small quantities for relatively minor attention. By Spring 2006, the works site in Loughborough was rapidly filling up with power cars and soon additional staff were recruited to deal with the amount of work required. FGW took the opportunity to have a substantial bodywork overhaul and cab refurbishment completed as part of the R41 installation. With several of its power cars entering the works following a protracted period out of service, the time taken to complete some vehicles was far longer than had been anticipated. No 43175 was the first of the 115 conversions for FGW to be completed and was outshopped on 7 July 2006.

The initial programme was completed when Nos 43002 and 43003 emerged on 6 March 2008. Once in full swing and with the most time-consuming power cars already

treated, Brush generally took eight to nine weeks per power car throughout 2007. The 115 power cars converted as part of the main programme for FGW were Nos 43002/003/005/010/012/015-018/020-037/040-042/063/069-071/078/079/086-088/091-094/097/098/122/124-156/158-165/168-172/174-177/179-183/185-198. This included the seven FGW power cars with VP185 engines, releasing those engines for further service elsewhere, and standardising the entire FGW fleet of 117 power cars with R41 engines coupled to Voith cooler groups. The two experimental power cars, Nos 43004 and 43009 revisited Brush for modifications (mainly to the cab desks) to bring them into line with the production batch. When Nos 43053 and 43056 were declared surplus they joined the FGW fleet and were dispatched to Brush for conversion, between December 2008 and April 2009, bringing them into line with the previous 117 power cars. It should be noted that Nos 43053 and 43056 had already received a revised electronics package while with a previous operator. The replacement of the electronics was not included within the scope of the FGW work and its fleet retained the original equipment and wheelslip control system.

Other fleets with R41 engines and the revised electronics were renumbered with 200 added to their original number. FGW Nos 43053 and 43056 were due to become 43253 and 43256 (and 43053 emerged with the new number), but both retained their original numbers with 43253 reverting back to 43053 prior to entering service.

With a much smaller fleet and less ability to

ABOVE Having completed its day in service by working the 18.00 from London King's Cross to Edinburgh on 18 September 2015, No 43208 *Lincolnshire Echo* (with No 43319 at the rear of the train), waits to run the short distance to Craigentinny depot for overnight servicing. *(James Trebinski)*

release too many power cars to conversion at any one time, the conversion of the GNER fleet was more gradual. When the first power cars to emerge for GNER were outshopped on 1 December 2006, it was decided to add 200 to the previous number on each of its power cars to distinguish them from unconverted examples. No 43100 was the first to appear, renumbered as 43300.

As well as receiving the R41 engine and a Brush cooler group each power car treated as part of the GNER programme received an upgraded electronics package including a new wheelslip protection system, which was vastly superior to the original design. The new electronics were originally trialled in Midland Mainline power car No 43048 (with a VP185 engine) and were designed to work with any engine, although it was not until the R41 'repower' programme that the improved electronics were brought into wider use. The success of this system in reducing wheel wear has allowed heavy exam periods to be extended as power cars are able to operate greater mileages between bogie changes. This has added between six and twelve months to bogie life so these are now changed every 30-36 months, dependant on duty cycle. This brings a huge maintenance cost saving to all fleets except Great Western where the original electronics were retained.

The control cubicle modification also included a new switch enabling the ability to isolate pairs of traction motors. The switch has three positions: 'All In', '1+2 Out' and '3+4 Out' and would be left in the 'All In' position unless a traction motor defect occurred. When the switch is operated, power notches 4 and 5 are isolated on the affected power car to prevent overloading the remaining traction motors. This is a useful feature that was omitted upon construction and has been subsequently fitted

ABOVE No 43004 is seen at Bristol Temple Meads on 29 August 2005 soon after entering service with the second trial-fit R41 engine. *(Chris Martin)*

BELOW No 43094 is seen in the yard at Brush undergoing testing during August 2006 prior to entering the paint booth. *(Chris Martin)*

RIGHT The last conversion for FGW is nearing completion as No 43056 is seen at Brush in April 2009. Note the orange junction boxes on the bogies. These are an indicator that this power car is fitted with the upgraded Brush electronics.
(Stuart Maclean)

to all fleets including the power cars with VP185 engines.

GNER decided bodywork overhauls would continue to be done when they became due, so this work was not generally included in the scope of the programme at Brush unless the power car was due for such attention. While the project was underway the parent company that owned GNER experienced financial difficulties and a new franchise for the ECML was awarded to National Express East Coast (NXEC) beginning in December 2007. However, this did not result in any interruption to the work underway at Brush.

Three additional power cars had been added to the programme by GNER, after the fleet was increased to facilitate service expansion. These were converted towards the end of the programme which was completed when Nos 43251 and 43310 (the former 43051 and 43110) emerged on 18 March 2009. The 30 power cars converted as part of the GNER-initiated programme became (after 200 had been added to their number) Nos 43206/208/238/239/ 251/257/277/290/295/ 296/299/300/302/ 305-320/367.

When a new franchise for CrossCountry (XC) began in November 2007 it was announced five IC125s would be added to its fleet following a period of several years since the type had been eradicated from the route by the introduction of new trains. Ten power cars would join the XC fleet and these would receive R41 engines, Brush cooler groups, new cab desks and the revised electronics package making them largely identical to the East Coast fleet.

Unusually, the earlier version of the MDEC control system panel without LCD screen was specified. As part of the attention at Brush Traction each CrossCountry power car was given a bodywork overhaul, all ten receiving the work during the latter half of 2008 with No 43366 being the last to emerge on 23 January

ABOVE Two power cars received nameplates connected to the MTU installation. No 43020 was named *mtu Power. Passion. Partnership.* while No 43290 (as illustrated) was christened *mtu fascination of power.*

BELOW Advances in electronics since the 1970s have made replacement of aging equipment worthwhile. This is the top of the new electronics rack as fitted to all power cars except those of FGW.

RIGHT The left-hand side of the bottom row of a new electronics rack.

FAR RIGHT The right-hand side of the bottom row of a new electronics rack.

2009. The ten power cars converted during this programme became (after 200 had been added to their number) Nos 43207/285/301/303/304/321/357/366/378/384.

The decision to convert the trio of Network Rail power cars was taken in mid-2008. They would certainly be needed for the foreseeable future and the small fleet was maintained by Edinburgh Craigentinny alongside an otherwise all-MTU-engined fleet. As only one could be released at a time, Nos 43013, 43014 and 43062 were treated in numerical order between March 2009 and January 2010. Each received the R41 engine with LCD display MDEC panel and upgraded electronics package, but all three had previously received the Brush cooler group and cab desk modifications.

Open-access operator Grand Central had continued using the original Paxman Valenta, but as part of a deal to extend its track access rights they committed to converting their power cars to R41 engines in order to improve reliability. The six Grand Central power cars began entering Brush Traction in April 2010 and left the site two at a time, with the last pair emerging on 1 April 2011. As well as the R41 engine with LCD display MDEC panel they received Brush cooler groups and the revised electronics package.

The opportunity was also taken to give each power car a bodywork overhaul and the new cab desk while they were at Brush. Although these power cars had always retained their original numbers since the installation of equipment fitted during 1987/88 to enable them to work as Driving Van Trailers (DVTs) in conjunction with electric locomotives, it was decided to give them new numbers in the 43/4 range. After the removal of the remaining residual equipment from the DVT programme the only technical difference

between the 43/2 and 43/4 series power cars was the conventional buffer beam on the front of the 43/4s. Nos 43065/067/068/080/084/123 became Nos 43465/467/468/480/484/423 during the overhaul process.

Two further power cars visited Brush Traction to receive the MTU R41 engine. In 2011, Nos 43072 and 43074 were transferred from the East Midlands Trains fleet to the East Coast fleet and to bring them in line with the rest of the East Coast power cars they received R41 engines. They already had Brush cooler groups and the electronics package and were attended to between July and March 2012, emerging as Nos 43272 and 43274 respectively. The engines installed in these two vehicles came from spare stock as tighter emissions regulations, which came into force during December 2010, resulted in it not being possible to purchase brand-new R41s owing to revised emissions limits. Over the full life of the 'repower' project, from the trial duo in 2005 to the completion of this additional pair, a total of 170 power cars received the MTU 16V4000 R41 engine. All had been converted by Brush Traction in Loughborough where the site had gone from being unknown in terms of IC125 power car attention, to become the first port of call for any substantial work.

Whilst the R41 performed well with a significantly reduced number of engine failures, the cost of maintaining the engines turned out to be considerably higher than for the previous engine types. In order to comply with the strict warranty conditions, a light 'overhaul' (termed QL1) is carried out every 1,000 hours of operation which includes a full oil and filter change. Dependent on operator, the R41 engine is given a half-life overhaul (classified as a QL3) after 12,500 hours of operation.

This is carried out by MTU's UK plant in East Grinstead, and is extended to 15,000 hours on operators with a lower duty cycle.

A full-life overhaul (classified as QL4), undertaken by the MTU remanufacturing centre in Magdeburg, Germany, is carried out after 25,000 hours of operation. As with the half-life overhaul, this is extended to 30,000 hours on lower duty cycles.

On all other engine types the operational power unit hours are calculated based on the time the vehicle is in traffic. With the R41 the MDEC records all the time the power unit is actually running, meaning that idling on depots accrues hours, thus further reducing the time between overhauls. When compared with the use of the lowest-bidding supplier to overhaul the Paxman Valenta, the cost of the 'main dealer' attention as well as the significant increase in oil change frequency, has raised the routine operating cost of the R41 considerably. This is partly offset by improved fuel consumption and the removal of a cost implication for the operator, following a significant engine failure.

Early issues with exhaust fires, occurring twice on No 43176 during late January and early February 2007, resulted in that power car having to return to Brush for a replacement engine. A small number of early fires were traced to unsecured electronic injector wiring in the rocker head cover becoming degraded and short-circuiting. This led to the injector becoming permanently stuck open, filling the turbo with unburnt fuel. A modification to the wiring in the rocker head has cured this problem, but further fires involving the R41 engine have occurred and these have been traced to two similar failure modes.

There have been occasions when an injector has become stuck open during engine starting and this cross fed to the other three cylinders in the group of four, filling them with unburnt diesel. The mechanical fuel pump continues to operate and the engine attempts to continue running on one cylinder until the accumulated unburnt diesel ignites, causing a fire in the exhaust box on top of the engine. Shutting down the engine in this situation does not stop the engine trying to run. The only solution to this problem is to remove the fuel supply by shutting off the fuel lift pump, but few operating staff would consider that action in a fire situation.

The most common cause of fires has been the failure of an oil seal on the turbocharger. There were also problems with the way the 'engine stop' signal to the engine was sent from the cab control. The wiring was such that the engine would run until it received a positive 'engine stop' signal. If there was a delay between a fire being discovered and the button being operated the engine would not receive the signal as the cable had been burnt though, so it would not then shut down. Eventually, the logic of the wiring was changed so that if the signal was removed the engine would shut down. Additionally, the wiring was placed in 'Nomex' sleeving to protect it from fire. Some subsequent changes have been made to controls so that the previous 'Fire Alarm Test' button on the cab desk has been replaced by a 'Local Engine Stop' button and the fire alarm test function has been moved to the electrical cubicle.

Other early issues with the R41 involved the oil filters becoming clogged up far sooner than

anticipated on some engines and for a period, the affected engines required an oil and filter change every 500 hours. Another early problem related to hotter-than-expected exhaust temperatures. The material used around the silencers was becoming degraded and a higher specification material was sourced to enable it to withstand higher temperatures.

Initially, the R41 used the standard power car fuel lift pump as fitted since construction, but it was decided in 2008 to modify the

old pump which had a pressure relief valve setting of 2.5bar, with a lower pressure relief valve setting of 1.3bar. This was more suitable for the R41 application and at the same time the opportunity was taken to improve the mounting and simplify the pipework arrangement.

New engine governors

After overcoming the early issues, the VP185 settled down to become a very capable engine in the IC125, but as a first-generation digital governor, the Viking Series 22 suffered from being reliant on old technology, and the unit itself was becoming obsolete with scarce availability of parts. As electronic control technology had advanced significantly, a far finer control of the engine would be possible, reducing fuel consumption and exhaust emissions. With insufficient serviceable governors to cover the fleet of 24 power cars, the decision was taken to trial a replacement Woodward digital governor. No 43089 was chosen to receive the experimental unit and this power car re-entered service on 5 December 2015.

The progress in technology has resulted in a far smaller box replacing the substantial unit mounted on the luggage van wall. As the new box is small enough to be held easily in one hand it was possible to place it in a custom-made frame mounted on top of the alternator, allowing the wiring between the control unit and the actuator on the engine to be minimised. An LCD panel on the governor enables status information to be displayed, which is supplemented by some LEDs and gauges permitting operation to be monitored easily by drivers and enabling laptop monitoring by engineers. The new governor also allows the possibility of a future upgrade to full electronic fuel injection of the VP185. Following the successful installation on No 43089, two further power cars (Nos 43043 and 43064) also now have Woodward governors.

The 'Hayabusa' Project

This really falls into the category of trialling new technology using IC125 vehicles, rather than an attempt to improve, enhance

or update the IC125 itself. Nevertheless, it was a remarkable experiment that warrants mention here. In a joint project between Hitachi, Porterbrook, Brush and Network Rail, No 43089 was substantially re-engineered, along with redundant trailer vehicle No 44062, to make a true 'hybrid' train with the ability to run using diesel or battery power. It was unveiled on 3 May 2007.

The batteries and control electronics were housed in No 44062 which was semi-permanently coupled to the power car, which was given three-phase AC traction motors. A substantial amount of wiring within and between the vehicles made the combination work as a hybrid. Three substantial, three-phase cables were run from the alternator into the trailer car where they reached a rectifier, creating a DC-link, which was also fed by 48 Lithium Ion batteries in the trailer vehicle. A standard Valenta engine and Brush alternator were still fitted and a Brush cooler group was installed for the trials.

The engine governor was specially modified to allow continuously variable air control of the speed. This meant the driver would select a power notch but the electronics would work out how the requested power would be delivered. Typically, a full-power start in notch 5 would see the engine remain at idle while the batteries provided traction power. As speed increased (along with the tractive effort required), the diesel engine would be brought 'on stream' by the control system and the speed and output increased to blend with

LEFT Forward-facing CCTV was fitted to selected power car fleets. This is the control box in an Abellio ScotRail power car luggage van.

the battery power. The system was intelligent enough to re-charge the batteries using power from the traction motors during braking. The name *Hayabusa* was applied to the power car, reminiscent of earlier prototype locomotives produced by Brush, Loughborough named after majestic birds. With Japanese involvement in this ground-breaking project, naming the power car after a Japanese peregrine falcon seemed most appropriate.

The test train was based on the Great Central Railway (GCR) for two weeks of trials before joining the New Measurement Train fleet, where it remained until September 2008. This enabled a huge amount of 'real world' data to be gathered on the new technology, which proved to be a huge success. Fuel efficiency figures were completely revolutionary for a diesel train, there was reduced stress on the diesel engine, and the traction equipment was virtually maintenance free. After the trial had been completed, No 43089 reverted back to standard configuration and returned to passenger service.

BELOW A special launch took place on 3 May 2007 to display hybrid power technology to the press. No 43089 *Hayabusa* is seen attached to modified coach No 44062 at Quorn and Woodhouse station on the Great Central Railway, with No 43160 *Porterbrook* at the rear of the train. awaiting departure with invited guests on board for the demonstration run to Rothley. *(Paul Biggs)*

Chapter Seven

Prototype returned to action

As the working lives of some of the fleet draws to a close, attention turns to preserving and restoring examples of the train. 125 Group is several steps ahead, having secured a pool of original Paxman Valenta power units and other major components. This chapter looks at their restoration of prototype No 41001 from a static museum exhibit to a fully working locomotive.

OPPOSITE Getting No 41001 back into action was just the start. Many more hours of volunteer time, effort and money have been spent keeping the power car operational. The 125 Group has also purchased some Mk3 carriages to recreate the authentic IC125 experience, albeit running at less heroic speeds on preserved lines! The first three Mk3s were repainted to match No 41001, and the train is seen at Loughborough on 24 February 2018 during one of the many occasions it has provided the service at its home railway, the GCR(N). *(John Zabernik)*

125 Group was formed in 1994 as an enthusiast organisation dedicated to reporting on the IC125 during its service career. There was also a long-term aspiration of building a foundation for possible future preservation of IC125 vehicles. When fleet-wide 'repower' ousted the original Paxman Valenta engines and Marston Excelsior cooler groups, a number of these items were secured by the group during 2007 and stored ready for future refitting to a power car, when they finally left service.

The demise of the original-configuration IC125 power car from normal service threatened to create a void in interest, but fortunately, discussions with the National Railway Museum (NRM) resulted in both parties wishing to see prototype power car No 41001 restored to working condition.

125 Group launched Project Miller on 14 May 2011 to undertake this considerable task. An arms-length division of the group would take No 41001 on loan, install a Paxman Valenta engine which belonged to the NRM, and undertake the rest of the work needed to restore the power car to operational condition after having been a static museum exhibit for nearly three decades.

LEFT Detailed examinations of No 41001 were carried out before the project got underway. The alternator is somewhat different from the production version but was found to be in good condition.

RIGHT Paxman Valenta engine No S508 was received many months before the power car itself. This engine was new in 2000 and had seen little service and had been last used in No 43143. Attention was needed to the SA-084 turbocharger, engine pipework and governor amongst other smaller items, before the engine would be ready to install.

RIGHT The magnificent generosity of East Midlands Trains in allowing the use of their depot at Neville Hill, Leeds (NL) as a base for the bulk of the restoration to be carried out cannot be overstated. No 41001 was transferred from the NRM to NL as an unbraked vehicle on 29 March 2012. Prior to the movement, considerable paperwork was needed to certify the vehicle as fit to move on the rail network for the first time in more than 25 years. No 41001 is seen moments after first arriving in the Back Fitting Shop at NL.
(Tony Shaw)

ABOVE One of the first tasks was removal of the main reservoir air tanks for overhaul and re-certification, which was very kindly done by Wabtec, Loughborough. This would allow No 41001 to be operated as an air-braked vehicle for all future movements.

BELOW The prototype's unique alternator was carefully mated to No S508 on 28 June 2012. The engine was lowered into No 41001 the following day.

ABOVE Following a brief spell back in the NRM for their 'Railfest' event from 29 May to 22 June 2012 No 41001 returned to NL and work continued in earnest. Valenta No S508 was delivered to NL and the partly sectioned No S183, which had been in No 41001, was removed ready for installation of the replacement engine.

RIGHT One of the necessary bodywork tasks was replacement of the corroded rear gangway connection. No 41001 is seen with this work in progress.

RIGHT Wiring in No 41001 was in a poor state, suffering from aged butyl rubber cables, disconnected items and undocumented modifications, so the decision was taken to lift the cubicle out, strip and refurbish it before replacing the wiring. The empty cubicle is seen awaiting re-wiring on 14 September 2012.

RIGHT Cooler group-to-engine hose connections proved a problem. The hoses are longer on the prototype compared with a production example, so affordable hoses had to be sourced and fitted.

ABOVE A Class 56 electronics rack and DC short circuiter were installed enabling the use of production power car type electronic modules, massively simplifying the task of fault finding and rectification. This is the electronics rack installed in the overhauled cubicle.

BELOW The cab desk during re-wiring. No photograph can do justice to the work needed to install nearly 2km of wiring and fitting all the necessary crimps and idents. Once complete, all the work is hidden away in cable ducting, behind panels and inside the cubicle itself.

BELOW The age and unique nature of the brake pads introduced a complication. Production power car pads would not fit without expensive modification, but fortunately Becorit stepped in and relined the original pads with new material. New tread brake blocks were also fitted all around the power car at this time, completing the braking system.

ABOVE The newly overhauled electrical cubicle was lowered back into No 41001 on 26 May 2013. Along with considerable other attention elsewhere on the electrical systems, the work undertaken turned a messy maze of dilapidated wiring into a reliably functioning system.

ABOVE The historic moment when No 41001's engine was started for the first time, outside the back fitting shop at NL early in the evening of 1 July 2013. Poor cylinder compression and dirty injectors had prevented this glorious moment being achieved earlier in the day. However, an old trick involving pouring engine oil into the liners freed things up enough to achieve this first start-up. *(Tony Shaw)*

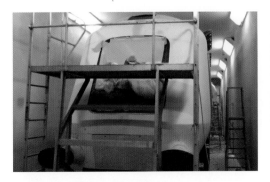

ABOVE Bodyside rot around the vents was tackled and the power car was rubbed down for repaint. No 41001 is seen in the confines of the NL paint booth on 11 July 2013. *(James Trebinski)*

ABOVE With repainting complete, No 41001 looked better than new when parked outside the Back Fitting shop at NL on 16 August 2013. The paint not only makes the power car look fantastic but will protect the vehicle from corrosion when kept outdoors during its new preservation career. *(Tony Shaw)*

LEFT It was not until the engine was started that another major hurdle was discovered. No 41001's prototype alternator was found to be an unmodified machine with an output of 75Hz making it incompatible with production-version three-phase fed electronics. After much head scratching and expert help from the electronics supremo at Brush Traction, a solution was found and a Class 57 traction control system was installed. One of the team is seen examining the alternator during the frustrating stage when nothing appeared to work as expected. Wiring on the alternator was later renewed with modern insulated cable replacing life-expired high-impedance wire.

ABOVE The hospitality at NL could not last for ever, so after 17 months in Leeds, during which invaluable help supplemented the 2,500 hours of volunteer effort, it was time to head to pastures new. The Great Central Railway Nottingham (GCRN) agreed to provide a new home for No 41001 at its Ruddington base, allowing restoration work to continue ahead of the power car becoming a 'home' loco on the line to Loughborough. The team worked well into the twilight and beyond on many occasions to complete the project. *(James Trebinski)*

ABOVE To complete the new electrical system a rectifier able to cope with auxiliary and 850v DC Electric Train Heating output was needed. Once again, consulting the experts fast-forwarded the team to a robust working solution and a new rectifier box was constructed and installed above the alternator. This is the new 'home-made' rectifier box with cover removed.

RIGHT Another problem – another solution. In order to regulate the output from the auxiliary supply an Automatic Voltage Regulator (AVR) would be needed. Rather than revive old and unreliable equipment the easiest solution was a new unit using MOSFET transistors. Fortunately, the team was put in contact with Noel Craigen and he used his experience in overcoming failures with aging equipment on Class 47s to produce a bespoke AVR box for No 41001. This was installed above the electrical cubicle in late 2013.

BELOW The new electronics rack was installed during March 2013 after a period of bench testing. The input board, output board, transducer interface board, traction control board, and an alternator control module were all slotted into place to complete the rack.

BELOW The compressor on No 41001 required attention and after toying with just changing the motors it was decided that an entire Class 73-type compressor would be fitted instead. Using a forklift, brute force and determination, the job was done on 30 May 2014.

ABOVE The biggest landmark during Project Miller was reached on 31 May 2014 when No 41001 was operated under its own power for the first time in 32 years. Some low speed, low power trials around the yard at Ruddington were successful, so a short train was formed and the momentous journey commenced. A brief silencer fire dealt with 30 years of accumulated dust and oil. Following that the test run was a complete triumph. The successful Project Miller team are seen at Loughborough after arrival, having reached the end of the GCR(N) line, powered by No 41001. *(James Trebinski)*

ABOVE RIGHT The original brake pipe pressure control unit (E70) had suffered from decades of inactivity and repairing it would have been problematic, and only delaying an inevitable failure. Using modern electronics and a new pressure transducer, a new E70 was built and installed in place of the original.

BELOW Further test runs allowed the team to iron out fuel starvation issues, adjust the governor and resolve intermittent power problems. The old and fragile silencer was changed and a decision was taken to replace the SA-084 turbocharger with the newer, NA256 version, as the original was passing oil. This is the second test run, paused in Fox Hill Cutting on 29 July 2014 while the team took a short break. *(John Zabernik)*

BELOW No 41001's big public launch came on 15 November 2014 when the first passenger-carrying train was operated. Christened 'The Screaming Valenta', a shortened six-car set was provided by East Midlands Trains and ran from Derby to Ruddington where No 43045 was removed and replaced by No 41001 for a run to Loughborough and back. No 43054 was on the other end of the train. Invited guests and 125 Group members alike were treated to a fault-free journey hauled by the power car, the first passengers to have done so since 1976. No 41001 is understandably the centre of attention as it waits to depart from Ruddington on this historic journey. Nose-cone designer and 125 Group Honorary President Sir Kenneth Grange is seen posing in front of his creation. *(Paul Colbeck)*

Technical specification

Production IC125 Power Car

General Information

Built:	1976-82 at BREL, Crewe
Maximum speed:	125mph
Wheel arrangement:	Bo-Bo
Min curve negotiable:	90.53m (4.5 chains)
Weight (as built):	Approx. 66t (drained), 70.25t (service ready)
Route Availability index:	5
Maximum tractive effort:	80kN (17,980lb)
Continuous tractive effort:	46kN (10,340lb)
Power at rail:	1,320kW (1,770hp)
Brake force:	35 tonnes
Fuel tank capacity:	4,680 litres (1,029gal)
Coolant capacity:	682 litres (150 gal)
Luggage van load limit:	43002-152 1.5 tonnes
(as built)	43153-198 2.5 tonnes

Dimensions

Length over headstocks:	17,364mm
Length over gangway:	17,792mm
Height:	3,810mm (over roof sheets), 3,906mm (max height)
Width over body:	2,740mm
Wheelbase:	12,900mm
Bogie wheelbase:	2,600mm
Bogie pivot centres:	10,300mm
Wheel diameter (new):	1,020mm

Equipment Specifications

Main alternator type:	Brush BA1001B, modified to BA1001C on MTU-engined power cars
Main alternator rating:	1,868kW
Auxiliary alternator type:	Brush BAH601B
Auxiliary alternator rating:	450kW
Auxiliary output range:	276V/33.3Hz to 415V/50Hz
Cooler Group (as built):	Marston Excelsior (43002-152)
	Serck Behr (43153-198)
Bogie type:	BP10B (Brush traction motors)
	BP10C (GEC traction motors)
Traction motors:	4 x Brush TMH68-46 (43002-123/153-198)
	4 x GEC G417AZ (43124-152)
Gear ratio:	59:23

Engine:

Engine Type	Paxman Valenta RP200L	Mirrlees Blackstone MB190	Paxman/MAN VP185	MTU 16V4000 R41
Configuration	60 degree V	45 degree V	90 degree V	90 degree V
Cylinders	12	12	12	16
Control	Regulateurs Europa 1122/2G Electro-Hydraulic Governor [1]	Heinzmann Electronic Governor	Regulateurs Europa Viking Series 22 Electronic Governor [3]	MDEC MTU Diesel Electronic Control (incorporates cooler group control)
Displacement	79 litres	70 litres	63.2 litres	65 litres
Max output (IC125 power car application)	1,678kW (2,250hp)	1,790kW (2,400hp)	1,678kW (2,250hp)	1,678kW (2,250hp)
Turbocharger(s)	1 x Napier SA-084 [2]	1 x Brown Boveri	6 x Schwitzer	4 x MTU
Number of power cars to have operated with type	197	4	35	170

Notes:
[1] Some engines fitted with Regulateurs Europa 1122/3G Electro-Hydraulic Governor
[2] Some engines modified with a Brown Boveri or Napier NA256 turbocharger
[3] Some power cars modified with Woodward electronic governor

Engine types fitted to each power car (original vehicle numbers used):
Paxman Valenta: 43002-198
Mirrlees MB190: 43167-170
Paxman/MAN VP185: 43043-050/052/054/055/058-061/064/066/072-076/081-083/089/165/167-170/173/177/179/191
MTU 16V4000 R41: 43002-010/012-018/020-042/051/053/056/057/062/063/065/067-072/074/077-080/084-088/090-172/174-198

Build differences by batch

	43002-055	43056-123	43124-152	43153-190	43191-198
Lot No	30876	30895	30941	30946	30968
Ordered	22 Jan 1974	24 Dec 1974	4 Apr 1978	1 Dec 1978	24 Jun 1980
Design code	GB502 0A	GB502 0A	GB502 0B	GB503 0A	GB503 0A
Guard's compartment?	Yes	Yes	Yes	No	No
Luggage van capacity	1.5 tonnes	1.5 tonnes	1.5 tonnes	2.5 tonnes	2.5 tonnes
Cooler Group	Marston Excelsior	Marston Excelsior	Marston Excelsior	Serck Behr	Serck Behr
Bogie type	BP10B	BP10B	BP10C	BP10B	BP10B
Traction Motors	Brush TMH68-46	Brush TMH68-46	GEC G417AZ	Brush TMH68-46	Brush TMH68-46
Rectifier	36 diode	36 diode	36 diode	12 diode	12 diode
Brake pipe pressure control unit	Davies & Metcalfe E70	Davies & Metcalfe E70	Davies & Metcalfe E70	Westinghouse DW2	Westinghouse DW2
Wheel slide protection system	BR Mark I	BR Mark II	BR Mark II	BR Mark II	BR Mark II
Parking brake equipment location	Guard's comp	Guard's comp	Guard's comp	Clean air	Clean air
Compressor	Davies & Metcalfe	Davies & Metcalfe	Davies & Metcalfe	Westinghouse	Westinghouse

Power Car Liveries

The prototype HSDT set was turned out in the "Pullman" version of British Rail's blue and grey livery scheme which simply reversed the standardised colour application making light grey the main colour with dark blue applied around the window line. For the production version of the Inter-City 125 it was decided that the standard blue and grey livery would be used for the carriages but power cars would wear a black and yellow scheme forward of the guards van area distinguishing them from the rest of the train. The first three power cars were turned out in this style before it was decided that the black would be replaced by blue and the initial trio were partially repainted before entry into service resulting in a uniform appearance across all 197 production power cars. British Rail changed the livery for its Inter-City fleet in 1983 and again in 1987; repaints were generally carried out on power cars when they next became due for overhaul so it wasn't until 1992 when the full fleet of 197 power cars regained a uniform appearance with all now wearing the popular *INTERCITY* "Swallow" livery. In September 1996 the first of the privatised companies applied its own livery scheme and since then a wide

array of different liveries have been applied as trains and companies have changed hands or companies have decided to change their fleet livery. There have been numerous smaller variations to livery schemes mainly altering the style of application of the compulsory yellow on the cab front, changing the colour of cab roofs or the application, variation or removal of bodyside branding. There are far too many such differences to detail here but one that stands out as worthy of mention was the novel treatment of 43028; this was part repainted receiving an orange and grey cab during 1997 in the style of the Republic of Ireland's Córas Iompair Éireann. Two power cars received partial repaints to trial new livery schemes and these are also not detailed in the list below; 43167 received Great Western green/ivory on the A-side and front only in 1994 and 43194 received a darker version of First Great Western's short-lived "dynamic lights" scheme on the B-side and front in 2006 although neither power car operated in service in this condition. During 2016 43002 and 43185 were returned to original blue/yellow and *INTERCITY* "swallow" livery respectively to mark the 40th anniversary of IC125 introduction.

Over the life of the fleet the following livery schemes were worn by the power cars listed. Data is correct at the end of 2018, but with planned future re-livery for GWR and ScotRail power cars included.

- British Rail blue/yellow/grey: 43002-198
- Inter-City "executive" (first version with yellow over grilles): 43125/126/129/130/151
- InterCity "executive": 43002-089/091-101/103-170/172-198
- *INTERCITY* "swallow": 43002-198
- Great Western green/ivory "Merlin": 43002-005/008-012/015-028/030-037/040-042/124-152/163-165/ 168-179/181-183/185-192
- GNER blue/vermillion*: 43006/008/038/039/051/053/056/057/067/077/078/080/090/095/096/099/ 100/102/105-120/167/197
- Midland Mainline teal/tangerine: 43043-061/064/066/072-077/081-083/085
- Virgin Trains red/dark grey: 43006-008/013/014/062/063/065/067-071/078-080/084/086-094/097-104/ 121-123/153-162/166/178/180/184/193-198
- First Great Western green/gold "Fag-Pack": 43002-005/009-012/015-028/030-037/040-042/124-152/163-165/ 168-172/174-177/179/181-183/185-192
- New Measurement Train yellow: 43013/014/062/067/089/154/196
- First Great Western blue/pink/white "Barbie": 43002-005/009/010/012/015-037/040-042/063/071/079/086/088/091/ 124-152/156/161-165/168-172/174-177/179-183/185-192/195/196
- Midland Mainline blue/grey/white "Ocean blue": 43007/043-061/063/064/066/069-079/081-083/085- 089/091/104/156/159/161/162/166/178/180/184/193/195-198
- Cotswold Rail silver: 43070
- Hornby advert red/yellow: 43087
- First Great Western blue/purple "Dynamic lights": 43004/009
- First Great Western plain blue: 43002-005/009/010/012/015-018/020-037/040-042/053/056/063/069-071/078/ 079/086-088/091-094/097/098/122/124-156/158-165/168-172/174-177/179-183/185-198
- Grand Central black: 43065/067/068/080/084/123
- East Midlands Trains blue/orange: 43043-050/052/054/055/058-061/064/066/072/073/075/076/081-083/089
- National Express East Coast white/grey*: 43006/008/038/039/051/057/077/090/095/096/099/100/102/ 105-120/167
- Cross Country silver/plum: 43207/285/301/303/304/321/357/366/378/384
- East Coast grey/purple: 43206/208/238/239/251/257/272/274/277/290/295/296/299/300/302/305-319/367
- Grand Central black/orange: 43423/465/467/468/480/484
- First Great Western advertising for external bodies (various): 43012/126/144/146/148/163/186/192
- Virgin East Coast red/white: 43206/208/239/251/257/272/274/277/290/295/296/299/300/302/ 305-320/367
- National Railway Museum advert: 43238
- World War I commemorative: 43172
- Queen Elizabeth II Diamond Jubilee: 43186
- Queen Elizabeth II 90th birthday: 43027
- Great Western Railway green: 43004/005/016/040-042/092-094/097/098/122/153-155/158/170/186-189/192/ 194/198
- ScotRail "Inter7City" blue/white: 43003/012/015/021/026/028/030-037/124-152/163/164/168/169/ 175-177/179/181-183
- East Midlands Trains darker blue: 43423/465/467/468/480/484

*Power cars in the fleets of GNER (and later National Express East Coast) were renumbered during the MTU 'repower' programme, only the original numbers are shown in these sections of the list for clarity. During 'repower' the fleets of GNER (and later National Express East Coast, East Coast, Virgin Trains East Coast and LNER), Cross Country and Grand Central were renumbered, see appendix 3 for details.

Power Car names and dates

The following table details the numbering, date of entry into passenger service and the various names which have been carried by the IC125 power car fleet.

During the 'repower' programme 48 power cars were renumbered, initially only the power cars operated by GNER (and later National Express East Coast and East Coast) received new numbers with 200 being added to their original number. Cross Country followed suit with their ten 'repower' conversions and when the six Grand Central power cars were converted they were renumbered into the 43/4 series. First Great Western power car 43053 was renumbered to 43253 upon release from Brush, Loughborough in April 2009 with the intention being that both 43053 and 43056 would be renumbered to distinguish them from the rest of the First Great Western fleet owing to their upgraded electronics cubicles. The re-numbering plan was aborted and 43253 was renumbered back to 43053 prior to re-entering passenger service.

The table shows the date of entry into passenger service; generally power cars were delivered, or at least accepted onto the rolling stock library, between two and four months before actually conveying passengers. However, as many of the earlier deliveries for both the Western and Eastern Region carried out several months of testing and crew training prior to operating services for fare-paying passengers, the dates in this table give a more accurate picture of when the power cars began the task they were built for. It should be noted that 43060/061 entered service several months before their counterparts as they were loaned to the Western Region prior to being needed for East Coast Main Line services. 43120/121 are actually the 62nd and 63rd power cars built, the need for spare power cars had quickly become apparent and four such vehicles were ordered as part of the second batch. Rather than wait until this batch had been completed the two spare power cars intended for the Western Region were delivered out of sequence.

The days when locomotives had a 'name for life' were consigned to history during the IC125 era and as can be seen from the list many power cars have worn two, three or even four names over the years, whilst some have never received a name. The period during which the power car wore each name is shown, where a previous name shows only one month this indicates that this title was worn for only a short period of time. Almost all of the names which were applied before mid-1987 were removed when the power cars were next repainted, the original cast nameplate style (and indeed the original nameplate position mid-way along the bodyside) had fallen out of favour and new namings received a revised style of nameplate now located between the radiator grille and luggage van door. Most, but by no means all, of the earlier names did eventually reappear and where there was a gap between nameplates being re-applied this is shown in the table. For example 43002 was named *'Top of the Pops'* from August 1984 until June 1988, lost the name upon repaint but regained the same name in June 1991 with the newer style of nameplates before finally losing it again in April 1996. Each name is shown in the exact mix of upper and lower case lettering as worn on the *main* power car nameplate, any wording which only appeared on subsidiary plaques or 'plates has been omitted. Several names (generally those on power cars operated by GNER and names prefixed "Rio") were applied as stickers and as such had a short life expectancy. Data in this table is correct to the end of 2018.

Number	Renumber (Date)	Date into passenger service	Current Name (Date applied)	Previous Names (Date applied-removed)
43002		08/1976	*Sir Kenneth Grange* (05/2016-)	*Top of the Pops* (08/1984-06/1988 and 06/1991-04/1996) *TECHNIQUEST* (05/1998-01/2008)
43003		08/1976		*ISAMBARD KINGDOM BRUNEL* (04/2006-04/2018)
43004		09/1976		*Swan Hunter* (11/1990-10/1994) *Borough of Swindon* (05/1997-03/2004) **First for the future* **First ar gyfer y dyfodol* (07/2005-11/2016)
43005		09/1976		
43006	43206 (01/2008)	08/1976		*Kingdom of Fife* (06/2003-11/2007)
43007	43207 (12/2008)	08/1976		
43008	43208 (10/2007)	09/1976	*Lincolnshire Echo* (05/2011-)	*City of Aberdeen* (07/2003-12/2007)
43009		09/1976		*First transforming travel* (06/2005-07/2014)
43010		08/1976		*TSW Today* (10/1990-01/1993)
43011		09/1976		*Reader 125* (06/1992-10/1999)
43012		08/1976		*Exeter Panel Signal Box 21st Anniversary 2009* (06/2016-12/2017)
43013		10/1976	*Mark Carne CBE* (07/2018-)	*University of Bristol* (10/1986-05/1988) *CROSSCOUNTRY VOYAGER* (02/1996-10/1998)
43014		08/1976	*The Railway Observer* (06/2014-)	

LEFT The original style of cast nameplate was applied mid-way along the bodyside and featured smaller than standard lettering with crests (where applicable) usually incorporated into the nameplate. This is the original nameplate applied to No 43142.

BELOW LEFT Initially, the new style of nameplate to match the INTERCITY 'Swallow' livery featured silver lettering on a grey background. This is the nameplate applied to No 43103.

BELOW The nameplate style changed again to the reflective type, with grey lettering. Photographing these nameplates could be challenging, as the shiny surface acted like a mirror reflecting everything in the vicinity. This is the nameplate applied to No 43016.

Number	Renumber (Date)	Date into passenger service	Current Name (Date applied)	Previous Names (Date applied-removed)
43015		10/1976		
43016		09/1976		*G yl Gerddi Cymru 1992 Garden Festival Wales 1992* (04/1992-08/1994) *Peninsula Medical School* (08/2002-11/2007)
43017		09/1976	*Hannahs discoverhannahs.org* (06/2014-)	*HTV West* (05/1987-05/1989)
43018		09/1976		*The Red Cross* (09/1997-02/2007)
43019		09/1976		*Dinas Abertawe City of Swansea* (05/1987-06/1989 and 03/1991-11/2004)
43020		08/1976	*mtu Power. Passion. Partnership.* (06/2011-)	*John Grooms* (08/1993-08/2007)
43021		09/1976		*David Austin - Cartoonist* (01/2007-08/2017)
43022		10/1976	*The Duke of Edinburgh's Award Diamond Anniversary 1956-2016* (03/2016-)	
43023		10/1976	*SQN LDR HAROLD STARR – ONE OF THE FEW* (09/2015-)	*County of Cornwall* (11/1989-12/2006)
43024		10/1976	*Great Western Society 1961-2011 Didcot Railway Centre* (09/2011-)	
43025		10/1976	*The Institution of Railway Operators 2000-2010 TEN YEARS PROMOTING OPERATIONAL EXCELLENCE* (04/2010-)	*Exeter* (04/1994-09/2006)
43026		10/1976		*City of Westminster* (05/1985-06/1990 and 02/1991-07/2006) *Michael Eavis* (04/2015-05/2018)
43027		10/1976		*Westminster Abbey* (05/1985-05/1990) *Glorious Devon* (04/1994-02/2007 and 11/2008-06/2016)
43028		10/1976		
43029		11/1976		
43030		10/1976		*Christian Lewis Trust* (06/2000-11/2018)
43031		11/1976		
43032		12/1976		*The Royal Regiment of Wales* (12/1989-09/2007)
43033		12/1976		*Driver Brian Cooper 15 June 1947 - 5 October 1999* (08/2003-10/2017)
43034		12/1976		*The Black Horse* (10/1994-12/2007) *TravelWatch SouthWest* (03/2008-11/2018)
43035		12/1976		
43036		01/1977		
43037		01/1977		*PENYDARREN 2004 RAIL BICENTENARY* (02/2004-11/2017)
43038	43238 (08/2007)	02/1977	*National Railway Museum 40 Years 1975-2015* (09/2015-)	*National Railway Museum The First 10 Years 1975-1985* (09/1985-11/1989 and 02/1991-03/1997) *City of Dundee* (12/2003-12/2007)
43039	43239 (06/2008)	02/1977		*The Royal Dragoon Guards* (03/2002-08/2006 and 12/2007-02/2008)

![DERBYSHIRE FIRST nameplate]

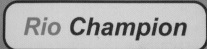

Rio Champion

IN SUPPORT OF HELP *for* HEROES

ABOVE After privatisation, Midland Mainline introduced their own unique style of nameplate featuring all-capital lettering and rounded corners. This is the nameplate applied to No 43082.

CENTRE A set of names was applied using a self-adhesive transfer towards the end of the temporary St Pancras-to-Manchester service, dubbed 'Project Rio'. This is the name applied to No 43156.

LEFT Bespoke nameplates with lettering in the style of the organisation they are recognising have been produced in some cases. This is the nameplate applied to No 43076.

Number	Renumber (Date)	Date into passenger service	Current Name (Date applied)	Previous Names (Date applied-removed)
43040		02/1977		*Granite City* (06/1990-12/1995) *Bristol St. Philip's Marsh* (01/2003-08/2018)
43041		02/1977	*Meningitis Trust Support for Life* (07/2011-)	*City of Discovery* (06/1990-10/2007)
43042		03/1977		
43043		03/1977		*LEICESTERSHIRE COUNTY CRICKET CLUB* (05/1997-11/2005)
43044		05/1977		*Borough of Kettering* (10/1993-10/2004)
43045		05/1977		*The Grammar School Doncaster AD 1350* (11/1983-02/1990 and 06/1991-12/1997)
43046		05/1977		*Royal Philharmonic* (10/1994-01/2006)
43047		05/1977		*Rotherham Enterprise* (03/1984-06/1989 and 02/1991-11/1997)
43048		04/1977	*T.C.B. Miller MBE* (04/2008-)	
43049		04/1977	*Neville Hill* (01/1984-)	
43050		06/1977		
43051	43251 (03/2009)	05/1977		*The Duke and Duchess of York* (07/1987-01/1989 and 06/1991-04/1998)
43052		05/1977		*City of Peterborough* (05/1984-10/1990 and 03/1991-01/1998)
43053	43253 (04/2009) 43053 (04/2009)	06/1977	*University of Worcester* (03/2010-)	*County of Humberside* (04/1984-07/1990 and 02/1991-04/1994) *Leeds United* (04/1994-10/2004)
43054		07/1977		
43055		07/1977	*The Sheffield Star 125 Years* (06/2012-)	*Sheffield Star* (11/1993-09/2005)

Number	Renumber (Date)	Date into passenger service	Current Name (Date applied)	Previous Names (Date applied-removed)
43056		05/1978	*The Royal British Legion* (04/2010-)	*University of Bradford* (11/1983-12/1990 and 12/1992-10/1997)
43057	43257 (12/2008)	03/1978	*Bounds Green* (03/1984-10/1989 and 08/1991-02/1998 and 06/2017-)	
43058		05/1978		*MIDLAND PRIDE* (02/1997-08/2003)
43059		04/1978		
43060		09/1977		*County of Leicestershire* (03/1985-04/1990 and 02/1991-01/2006)
43061		09/1977		*City of Lincoln* (05/1984-05/1988 and 03/1991-12/1997) *The Fearless Foxes* (10/2016-11/2018)
43062		02/1978	*John Armitt* (07/2007-)	
43063		02/1978		*Maiden Voyager* (01/1997-08/1999 and 07/2000-11/2003) *Rio Challenger* (08/2004-09/2004)
43064		05/1978		*City of York* (09/1983-10/1989 and 03/1991-02/1998) *125 Group* (07/2008)
43065	43465 (12/2010)	05/1978		*City of Edinburgh* (02/1996-11/1998)
43066		04/1978		*Nottingham Playhouse* (08/1995-04/2005)
43067	43467 (12/2010)	04/1978	**British Transport Police Nottingham* **Nottinghamshire Fire and Rescue Service* (02/2018-)	
43068	43468 (09/2010)	05/1978		*The Red Nose* (03/1997-03/1999) *The Red Arrows* (11/2000-05/2004)
43069		05/1978		*Rio Enterprise* (08/2004-05/2007)
43070		06/1978	*The Corps of Royal Electrical and Mechanical Engineers* (10/2007-)	*Rio Pathfinder* (08/2004-09/2005)
43071		06/1978		*Forward Birmingham* (09/1996-05/2003)
43072	43272 (11/2011)	05/1978		*Derby Etches Park* (07/1993-09/2005 and 05/2006-07/2011)
43073		06/1978		
43074	43274 (02/2012)	05/1978	*Spirit of Sunderland* (12/2015-)	*BBC EAST MIDLANDS TODAY* (05/1997-01/2003)
43075		05/1978		
43076		05/1978	*IN SUPPORT OF HELP for HEROES* (05/2010-)	*BBC East Midlands Today* (01/1991-03/1997) *THE MASTER CUTLER 1947-1997* (10/1997-05/2005)
43077	43277 (10/2008)	05/1978		*County of Nottingham* (09/1984-05/1989 and 05/1991-05/1997)
43078		05/1978		*Shildon County Durham* (09/1983-06/1990 and 03/1993-06/1996) *Golowan Festival Penzance* (06/1996-08/2003) *Rio Crusader* (09/2004)
43079		04/1978		*Rio Venturer* (08/2004-09/2004)

Number	Renumber (Date)	Date into passenger service	Current Name (Date applied)	Previous Names (Date applied-removed)
43080	43480 (09/2010)	07/1978	**West Hampstead PSB** (12/2018-)	
43081		10/1978		*Midland Valenta* (07/2008)
43082		05/1978	**RAILWAY children – Fighting for street children** (07/2017-)	*DERBYSHIRE FIRST* (11/1997-10/2004) *RAILWAY children – THE VOICE FOR STREET CHILDREN WORLDWIDE* (09/2009-07/2017)
43083		07/1978		
43084	43484 (04/2011)	10/1978		*County of Derbyshire* (07/1986-05/2004) *PETER FOX 1942-2011 PLATFORM 5* (05/2011-12/2017)
43085	43285 (08/2008)	06/1978		*City of Bradford* (06/1983-11/1989 and 04/1991-09/1997)
43086		07/1978		*Rio Talisman* (08/2004-07/2007)
43087		07/1978	**11 Explosive Ordnance Disposal Regiment Royal Logistic Corps** (11/2010-)	*Rio Invader* (08/2004-09/2005)
43088		06/1978		*XIII Commonwealth Games Scotland 1986* (03/1985-02/1989 and 03/1991-11/1999) *Rio Campaigner* (09/2004)
43089		09/1978		*Rio Thunderer* (09/2004 *HAYABUSA* (05/2007-09/2008)
43090	43290 (12/2006)	07/1978	**mtu fascination of power** (06/2007-)	
43091		07/1978		*Edinburgh Military Tattoo* (08/1985-11/1988 and 03/1993-02/1999)
43092		08/1978		*Highland Chieftain* (05/1984-01/1988) *Highland Chieftan* (03/1993-04/1993) *Institution of Mechanical Engineers 150th Anniversary 1847-1997* (04/1997-05/2004)
43093		08/1978	**Old Oak Common HST Depot 1976-2018** (09/2017-)	*York Festival '88* (02/1988-01/1997) *Lady in Red* (02/1997-09/2003)
43094		08/1978		
43095	43295 (12/2008)	09/1978		*Heaton* (02/1984-03/1991) *Perth* (11/2003-10/2008)
43096	43296 (07/2007)	09/1978		*The Queen's Own Hussars* (05/1985-12/1987 and 12/1992-03/1997) *The Great Racer* (05/1997-12/2000) *Stirling Castle* (11/2002-06/2009)
43097		09/1978	**Environment Agency** (09/2006-)	*The Light Infantry* (11/1983-04/1990)
43098		10/1978		*Tyne and Wear Metropolitan County* (09/1985-12/1987) *railwaychildren* (05/1998-05/2004)
43099	43299 (06/2008)	09/1978		*Diocese of Newcastle* (05/2006-10/2006)
43100	43300 (11/2006)	11/1978	**Craigentinny** (03/1984-12/1987 and 06/1991-03/1998 and 12/2006-)	*Blackpool Rock* (05/1998-05/2000)

Number	Renumber (Date)	Date into passenger service	Current Name (Date applied)	Previous Names (Date applied-removed)
43101	43301 (07/2008)	11/1978		*Edinburgh International Festival* (08/1984-10/1987 and 08/1993-07/1998) *The Irish Mail Tren Post Gwyddelig* (07/1998-07/2000)
43102	43302 (04/2008)	11/1978	*World Speed Record - HST* (11/2017-)	*City of Wakefield* (06/1984-10/1987) *HST Silver Jubilee* (12/2001-10/2003) *Diocese of Newcastle* (12/2006-01/2008)
43103	43303 (07/2008)	11/1978		*John Wesley* (05/1988-05/1998) *Helston Furry Dance* (05/2002-09/2003)
43104	43304 (12/2008)	11/1978		*County of Cleveland* (04/1985-02/1990 and 12/1992-01/2001) *City of Edinburgh* (03/2001-10/2002)
43105	43305 (08/2008)	11/1978		*Hartlepool* (07/1984-03/1988 and 01/1993-12/1996) *City of Inverness* (09/2001-12/2007)
43106	43306 (03/2007)	01/1979		*Songs of Praise* (06/1989-02/1997) *Fountains Abbey* (08/2003-12/2007)
43107	43307 (06/2008)	01/1979		*City of Derby* (05/1986-10/1988) *Tayside* (07/2003-03/2008)
43108	43308 (02/2009)	01/1979	*HIGHLAND CHIEFTAIN* (09/2014-)	*BBC Television Railwatch* (02/1989) *Old Course St Andrews* (07/2000-07/2002 and 11/2002-09/2008)
43109	43309 (06/2007)	02/1979		*Yorkshire Evening Press* (05/1989-03/1997) *SCONE PALACE* (09/2002-10/2002) *Leeds International Film Festival* (11/2005-12/2007) *Leeds International Festival* (12/2007-09/2008)

LEFT Prior to privatisation, Cross Country developed their own style of nameplates featuring italic lettering which was initially continued by Virgin Trains after they took over operation of the route. This is the nameplate applied to No 43158.

BELOW LEFT For three IC125 power car naming events Virgin Cross Country used large bodyside vinyl stickers instead of cast nameplates. This is the name applied to No 43100.

BELOW GNER applied names to most of their IC125 power cars using gold lettering rather than traditional cast nameplates, the departure from the more solid style of application often rendered the names short-lived affairs. This is the name applied to No 43108.

Old Course St Andrews

Number	Renumber (Date)	Date into passenger service	Current Name (Date applied)	Previous Names (Date applied-removed)
43110	43310 (03/2009)	03/1979		*Darlington* (05/1984-12/1988 and 02/1991-03/1997) *Stirlingshire* (03/2003-12/2008)
43111	43311 (08/2008)	03/1979		*Scone Palace* (10/2002-02/2007 and 09/2007-06/2008)
43112	43312 (03/2008)	04/1979		*Doncaster* (05/2001-12/2007)
43113	43313 (11/2007)	04/1979		*City of Newcastle Upon Tyne* (04/1983-02/1989 and 02/1993-04/1997) *The Highlands* (12/2003-12/2009)
43114	43314 (02/2007)	04/1979		*National Garden Festival Gateshead 1990* (08/1989-12/1996) *East Riding of Yorkshire* (01/2005-02/2009)
43115	43315 (02/2008)	04/1979		*Yorkshire Cricket Academy* (06/1989-07/1997) *Aberdeenshire* (02/2004-11/2007)
43116	43316 (10/2007)	05/1979		*City of Kingston Upon Hull* (05/1983-08/1989 and 06/1991-01/1997) *The Black Dyke Band* (12/2005-02/2008)
43117	43317 (04/2008)	05/1979		*Bonnie Prince Charlie* (08/2002-02/2008)
43118	43318 (05/2007)	05/1979		*Charles Wesley* (05/1988-11/1996) *City of Kingston Upon Hull* (11/2002-08/2009)
43119	43319 (11/2008)	05/1979		*Harrogate Spa* (05/2003-08/2008)
43120	43320 (01/2007)	09/1977		*National Galleries of Scotland* (11/2005-11/2008)
43121	43321 (09/2008)	09/1977		*West Yorkshire Metropolitan County* (09/1984-09/1988 and 03/1991-01/1999)
43122		05/1979		*South Yorkshire Metropolitan County* (01/1985-08/1989 and 02/1991-stolen whilst power car in storage sometime in 2004)
43123	43423 (04/2011)	05/1979	*'VALENTA' 1972-2010* (12/2010-01/2011 and 05/2011-)	
43124		06/1981		*BBC Points West* (09/1986-11/1989)
43125		06/1979		*Merchant Venturer* (04/1985-06/1989 and 06/1994-07/2006)
43126		06/1979		*City of Bristol* (04/1985-09/1987 and 09/1991-10/2006)
43127		07/1979		*Sir Peter Parker 1924-2002 Cotswold Line 150* (06/2003-03/2018)
43128		07/1979		
43129		08/1979		
43130		08/1979		*Sulis Minerva* (02/1992-10/2006)
43131		10/1979		*Sir Felix Pole* (08/1985-11/1987 and 04/1994-05/2007)
43132		10/1979	*Aberdeen Station 150th Anniversary* (11/2017-)	*Worshipful Company of Carmen* (10/1987-12/1995) *We Save the Children Will you?* (10/2007-08/2017)

Number	Renumber (Date)	Date into passenger service	Current Name (Date applied)	Previous Names (Date applied-removed)
43133		10/1979		
43134		10/1979		*County of Somerset* (07/1992-10/2007)
43135		10/1979		*Quaker Enterprise* (07/2005-11/2006)
43136		10/1979		
43137		06/1980		*Newton Abbot 150* (04/1997-10/2007 and 10/2010-11/2018)
43138		06/1980		
43139		04/1980		*Driver Stan Martin 25 June 1950 - 6 November 2004* (06/2005-08/2018)
43140		04/1980		**Landore Diesel Depot 1963 Celebrating 50 Years 2013* **Depo Diesel Gland r 1963 Dathlu 50 Mlynedd 2013* (11/2013-03/2018)
43141		05/1980		**Cardiff Panel Signal Box 1966-2016* **Blwch Signalau Panel Caerdydd 1966-2016* (04/2016-05/2018)
43142		05/1980		*St Mary's Hospital Paddington* (11/1986-03/1989) *Reading Panel Signal Box 1965-2010* (07/2011-04/2018)
43143		11/1981		*Stroud 700* (04/2004-11/2007 and 11/2008-11/2017)
43144		11/1981		
43145		03/1981		
43146		03/1981		
43147		05/1981		*Red Cross* (05/1988-02/1991) *The Red Cross* (02/1991-09/1997) *Royal Marines Celebrating 350 Years* (07/2014-10/2018)
43148		05/1981		
43149		06/1981		*BBC Wales Today* (09/1988-07/2007) *University of Plymouth* (03/2011-03/2018)
43150		06/1981		*Bristol Evening Post* (10/1988- 09/2007)
43151		06/1981		*Blue Peter II* (05/1987-11/1989)
43152		06/1981		*St Peter's School York AD 627* (11/1984-06/1989 and 06/1991-10/1995)
43153		01/1981		*University of Durham* (07/1983-07/1989 and 06/1991-02/1997) *THE ENGLISH RIVIERA TORQUAY PAIGNTON BRIXHAM* (06/1997-05/2004)
43154		01/1981		*INTERCITY* (03/1994-09/2003)
43155		01/1981	*The Red Arrows 50 Seasons of Excellence* (08/2014-)	*BBC Look North* (06/1985-04/1990 and 05/1991-04/1997) *The Red Arrows* (05/1997-06/1998) *City of Aberdeen* (06/1998-12/2004)
43156		01/1981	*Dartington International Summer School* (07/2007-)	*Rio Champion* (08/2004-09/2004)
43157	43357 (09/2008)	04/1981		*Yorkshire Evening Post* (01/1984-05/1990 and 05/1991-05/1998) *HMS Penzance* (07/1998-05/2000)

Number	Renumber (Date)	Date into passenger service	Current Name (Date applied)	Previous Names (Date applied-removed)
43158		04/1981		*Dartmoor The Pony Express* (03/1995-07/1998 and 04/2001-09/2003)
43159		04/1981		*Rio Warrior* (08/2004-02/2007)
43160		04/1981	*Sir Moir Lockhead OBE* (01/2011-)	*Storm Force* (04/1991-10/1997) *Porterbrook* (05/2007-08/2007)
43161		05/1981		*Reading Evening Post* (04/1991-05/1999) *Rio Monarch* (09/2004)
43162		05/1981	*Exeter Panel Signal Box 21st Anniversary 2009* (12/2017-)	*Borough of Stevenage* (03/1984-05/1999) *Project Rio* (08/2004-09/2004)
43163		11/1981		*Exeter Panel Signal Box 21st Anniversary 2009* (05/2009-06/2016)
43164		11/1981		
43165		10/1981	*Prince Michael of Kent* (10/2006-)	
43166	43366 (01/2009)	10/1981		
43167	43367 (04/2007)	12/1981	*DELTIC 50 1955-2005* (10/2005-)	
43168		12/1981		
43169		10/1981		*The National Trust* (07/1989-02/2018)
43170		10/1981		*Edward Paxman* (06/1995-10/2007)
43171		10/1981		

LEFT One of the most unusual nameplates to be worn by an IC125 power car was based on a much earlier style of nameplate applied to prototype locomotives in the 1950s and 1960s. This is the nameplate applied to 43089.

BELOW LEFT Very distinctive nameplates applied in later years included this title cast in the style used on the 1940s 'Battle of Britain' Bullied 'Pacific' steam locomotives which were built by for the Southern Region. The life of World War Two RAF pilot Harold Starr was commemorated by this nameplate applied to No 43023.

BELOW To complete a special World War One commemorative livery covering the entire power car bodyside, this nameplate was produced and applied to No 43172.

Number	Renumber (Date)	Date into passenger service	Current Name (Date applied)	Previous Names (Date applied-removed)
43172		10/1981	*Harry Patch – The last survivor of the trenches* (11/2015-)	
43173		10/1981		*Swansea University* (06/1996-09/1997)
43174		10/1981		*Bristol - Bordeaux* (04/1997-04/2007)
43175		11/1981		*GWR 175TH ANNIVERSARY* (09/2010-10/2018)
43176		11/1981		
43177		12/1981		*University of Exeter* (12/1995-12/2006)
43178	43378 (11/2008)	12/1981		
43179		12/1981		*Pride of Laira* (09/1991-12/2017)
43180		12/1981		*City of Newcastle Upon Tyne* (05/1998-05/2003) *Rio Glory* (08/2004-09/2004)
43181		12/1981		*Devonport Royal Dockyard 1693-1993* (11/1993-10/2007)
43182		12/1981		
43183		03/1982		
43184	43384 (11/2008)	03/1982		
43185		04/1982	*Great Western* (05/1992-)	
43186		04/1982		*Sir Francis Drake* (07/1988-05/2007)
43187		05/1982		
43188		05/1982		*City of Plymouth* (05/1986-08/1988 and 04/1990-02/2007)
43189		05/1982		*RAILWAY HERITAGE TRUST* (10/1995-11/2007 and 05/2010-07/2018)
43190		05/1982		
43191		06/1982		*Seahawk* (10/1988-07/2007)
43192		06/1982		*City of Truro* (10/1988-09/2007)
43193		09/1982		*Yorkshire Post* (02/1983-05/1992) *Plymouth SPIRIT OF DISCOVERY* (03/1995-06/2004) *Rio Triumph* (09/2004-07/2007)
43194		09/1982		*Royal Signals* (10/1985-12/1989)
43195		09/1982		*British Red Cross 125th Birthday 1995* (12/1995-07/2003) *Rio Swift* (08/2004-09/2004)
43196		09/1982		*The Newspaper Society* (04/1986-02/1989) *The Newspaper Society Founded 1836* (02/1991-02/2004) *Rio Prince* (08/2004-09/2005)
43197		09/1982		*Railway Magazine 1897 Centenary 1997* (11/1996-10/2000) *The RAILWAY MAGAZINE* (02/2001-10/2003) *Rio Princess* (08/2004-09/2004)
43198		09/1982	*Oxfordshire 2007* (02/2007-)	*HMS Penzance* (06/2000-06/2003) *Rio Victorious* (08/2004- 01/2006)

*Different name carried either side of power car

Appendix 4

Ownership changes

In April 1994 the entire British Rail passenger fleet was divided up into three rolling stock companies (ROSCOs) in preparation for privatisation. The IC125 fleet was split between two ROSCOs dependant on the route each vehicle was dedicated to work at that time. Subsequently some redundant vehicles were sold by Porterbrook to smaller ROSCOs which had been created by the holding companies of train operating companies. In order to finance the 'repower' programme the vehicles owned by Sovereign were sold to one of the major ROSCOs and then leased back to their operator. This table details the changes of power car ownership up to the end of 2018. Power cars withdrawn prior to this point had sustained collision damage beyond viable economic repair, during and beyond 2019 numerous power cars are expected to be withdrawn as replacement rolling stock renders them redundant.

Date	Ownership (change)
From new	British Rail: 43002-198
Apr 1994	Angel Trains: 43002-012/015-042/095/096/104-120/124-152/163-179/181-192 Porterbrook: 43013/014/043-094/097-103/121-123/153-162/180/193-198
Apr 1999	43173 withdrawn
Jul 2002	43011 withdrawn
Oct 2004	43068/084/092-094/097/098/122/123/153/155/198 Porterbrook > First Rail Holdings
Jan 2005	43019 withdrawn
Feb 2006	43154/158/194 Porterbrook > First Rail Holdings and 43068/084/ 123 First Rail Holdings > Porterbrook (swapped for standardisation)
Feb 2007	43065/067/068/080/084/123 Porterbrook > Sovereign
Feb 2010	43065/067/068/080/084/123 Sovereign > Angel Trains
May 2018	43195 Porterbrook > Great Western Railway (for spares recovery; accident damaged vehicle)

Index

During 2018 the next phase in the lives of the fleet began as new trains entered service displacing IC125s from Inter-City services operating from London Paddington, a route they have dominated since 1976. A less demanding role operating shortened formations on secondary routes will keep some of the trains in daily passenger operation into their 50th year and beyond. Some are remaining with Great Western (as depicted on page 13) whilst others are introducing Inter-City levels of comfort to ScotRail services. Newly overhauled carriages are fitted with sliding external doors to meet changes in disability access legislation which come into effect in 2020. Nos 43033 and 43183 are seen at North Queensferry having just crossed the world famous Forth Bridge whilst operating the 13.30 Edinburgh Waverley to Aberdeen on 18 October 2018. *(Chris Hopkins)*